WASTEnomics

WASTEnomics

edited by
Kenny Tang and
Jacob Yeoh

Middlesex
University
PRESS

First published in 2008 by Middlesex University Press

Copyright © Middlesex University Press

Authors retain copyright of individual chapters.

ISBN 978 1 904750 28 4

A CIP catalogue record for this book is available from

The British Library

Cover design by Helen Taylor

Typesetting by Carnegie Book Production, Lancaster

Middlesex University Press
North London Business Park
Oakleigh Road South
London N11 1QS

Tel: +44 (0)20 8411 4162: +44 (0)20 8411 4161
Fax: +44 (0)20 8411 4167

www.mupress.co.uk

Contents

Waste: the golden opportunity of our generation

Tan Sri Dr Francis Yeoh CBE

Group Managing Director, YTL Corporation

Waste – in any shape, form or mode – is the defining issue of our abundant generation. We produce more and more waste each year. Most of it comes from needlessly discarding and throwing away what we have spent enormous amounts of energy and precious resources in producing in the first place, most probably after just one use.

Put simply, waste is threatening the very survival of our planet – from waste that is causing air pollution to food pollution to water pollution. Waste is the root cause of many of society's problems today and in the future.

Why are we here at this juncture in time? The indiscriminate and unsustainable use of our planet's scarce resources, through an industrial world powered largely by dirty fossil fuels that emit billions of tonnes of greenhouse gases and toxic materials (such as mercury) into the air, land and water, is unsustainable in any shape or form for our planet and its citizens. Moreover, these resources are diminishing at a time when the world's population is growing exponentially.

Looking to the future, if we project that living standards and population double over the next 100 years, and if we assume that the developing countries of China, India, Brazil and Indonesia will share the same living standards as the developed world, our use of resources (and the attendant waste) will need to increase by a factor of 10 to 20 in the next century. It is quite impossible to imagine that we can increase industrial throughput by a factor anywhere near 10, let alone 20, considering the earth's dangerously depleted and damaged life-support systems – even without the dangerous effects of climate change.

Therefore, in addressing the problem of waste, businesses could address a number of strategic but inter-related issues at the same time. I have no doubt that the mega opportunities in the next 50 years will be the ones that address the issue of waste and its inter-related problems of energy and resource use. Can we simply and intelligently design goods that are 'earth-to-earth' (not a one-way traffic from 'earth-to-landfill')? Could we move away from the fossil-fuel-led carbon production machine of the past and the present while moving towards a sustainable route map to the future?

In tomorrow's resource-constrained world where the problems of waste will be seriously addressed, business as usual is a dangerous strategy. Corporate and global leaders urgently need to embrace the planet-saving message of our future resource-constrained world. Society today resides on an impressive platform of a global infrastructure built on a diet of fossil fuels and carbon. Beneath the carbon and resource footprint of the world's corporations, the foundations of the earth that they have taken for granted are disintegrating – including the dangerous effects of human-induced climate change and air, water and food pollution caused by a 'winner-take-all' industrial and capitalist system.

Visionary leaders realize that a major industrial revolution is inevitable – a road map is needed to cross the chasm to a new economy that relies less on resources, and a licence to operate that is not based on a mantra of 'take–use–discard'. A new strategic and industrial vision is necessary; but even these innovative leaders struggle to visualise what it could mean for them. Meanwhile, others have not yet even realized that they need an exit route out of our present diet of dirty fossil fuels.

What is the new vision? How can we reverse the trend? *WASTEnomics* is a unique book and a first step towards the impending industrial revolution that will re-shape our view of waste in our resource-constrained society.

This will only be a starting point. The impending revolution in our industrial and capitalist systems that will move us away from a diet of fossil fuels and unlimited resources requires vision, ingenuity, energy and creativity from all citizens of all nations. The future belongs to those who understand that doing more with less of our precious resources is common sense and compassionate, and thus more intelligent and competitive. The prize is a huge one for the sake of the planet. Visionary leaders must aim for nothing less.

Zero is not an option

John Elkington
Co-founder and Chief Entrepreneur at SustainAbility,
Co-founder of Volans Ventures

Ever-improving resource efficiency and waste management are not an option in today's world; but, for many business and community leaders, adopting the goal of 'zero waste' very definitely is. Various international companies may have bitten the bullet in recent years and adopted zero-based targets in areas like safety and emissions, among them DuPont and BP, but most business and community leaders – if they pay much attention at all to resource and efficiency and waste management – tend to shy away from anything that smells like ambition in this field. This is not an option they see as either commercially or politically viable.

Yes, the energy crisis created by oil priced at or around $100 a barrel will help on this front, but high energy prices also sometimes hit waste management and recycling hard, too. Ultimately, it is increasingly clear that the challenges we face in this area will not be met by doing more of the same. Recall the wisdom attributed to Einstein: 'the significant problems we have cannot be solved at the same level of thinking with which we created them.' The time has come to rethink not just how we handle, reclaim and reprocess waste materials, but how we design our economies, value chains, technologies and products in the first place. There has been much talk of the next Industrial Revolution, but that – in effect – is what now needs to happen.

One of the biggest shifts proposed in this book is that we should all come to own the waste we generate, rather than simply subcontracting the problem to others. Governments in some countries are already moving in this direction by imposing waste charges, or by imposing limits on how much waste households can generate, and a great deal more needs to be done on this front to ensure even a reasonable degree of progress. Business has a key role to play here, not just in terms of the obvious investments in more sustainable business models, technologies and products, but in terms of abandoning defensive positions on waste and, instead, lobbying in favour of effective, efficient controls that promote not only more sustainable forms of waste management but also a huge surge in the relevant areas of innovation and entrepreneurship.

The four principles of WASTEnomics presented, explained and promoted in this book certainly draw on best practice – but they also move well beyond, to signal some of the new directions that the twenty-first century economy will take us in. We can debate when and where the 'tipping point' will be achieved, but in a world headed towards nine or ten billion people and with consumerist lifestyles the default setting for pretty much all societies, the question is not now whether there will be a New Industrial Revolution, but when and where it will start, and who will benefit and who lose as it begins to build irresistible momentum.

Dedicated to our future generations

The Four Core Principles of WASTEnomics

- *Waste to **Zero***

- *Waste to **Own***

- *Waste as **Societal Liability***

- *Waste as **Asset***

Introduction and acknowledgements
Dr Kenny Tang CFA and Jacob Yeoh

There are limits to resources but none to human creativity (Thierry Volery)

Waste! Waste! Waste! In 2002, a staggering 32 billion cans of fizzy drinks were thrown away in America that, if recycled, would have saved 435,000 tonnes of aluminium – enough to rebuild the entire commercial air fleet of the world. A massive 45 million tonnes of containers and packaging are discarded annually in the US creating a host of unwanted and unintended environmental problems.

This is only the physical waste we can see and hold – what about the waste we cannot see such as greenhouse gases, toxic chemicals and radioactive wastes in the air, land and water?

Throughout recent history we have treated the planet's precious resources with almost utter disdain and taken them for granted as if such resources are unlimited. Whatever we take from the planet, we use and throw away – a one-way traffic to the sea or landfill or incinerator, sometimes after just a single use.

Most notably since the Industrial Revolution, our methods of mass production have been built on the mantra of 'take, use, throw' – where we take the planet's precious resources, use and make what we like out of them while emitting chemicals into the air and water, and throw away the unused inputs and unwanted outputs. Inevitably these unwanted inputs and outputs, past their use-by date, end up either in a landfill (in the case of computers and television sets with toxic and other materials such as lead and copper still embedded), in the sea, in the incinerator or – in the case of extractive resources – in contaminated land or water.

Tipping point moment

We are at a major tipping point moment in our planet's history as we discover that the quantities of waste and pollution in the air, land and water are in danger of quite literally suffocating us.[1] It is at this point that we fully realise the waste (and resource use) problem caused by our growing population, increasing city urbanisation and 'package-everything-with-plastics-and-throw-away' lifestyle. The problem will be insurmountable unless and until we change our resource-based consumption habits, our outdated and

---

1   For example, the recent high profile campaign (February 2008) by the *Independent* and the *Daily Mail* newspapers in the UK to reduce plastic pollution in the seas and to ban free plastic bags caused many to realise the quantity of such plastic bags (each UK individual uses an average of 216 plastic bags per year, making a total of 12 billion bags) and their durability.

flawed models of production and the indiscriminate and irresponsible way we extract our raw materials, as if the supply is unlimited. *Something has to change.*

Furthermore the oil price has reached the magic US$100 per barrel in 2008. Crude oil, as we know, has downstream impact in many industrial areas, including the cost of energy generation and motor transport. It is also the input into many products, including plastics. It is also a fact that any form of waste (be it complete goods or leftovers) has considerable value already embedded in it through the energy used either in its production or its transport.

The first Industrial Revolution, over two centuries ago, made human talent highly productive because people were relatively scarce while nature and its raw materials, at the time, seemed almost boundless. The plentiful supply of materials created new industries from the methods of mass production but, as a consequence, produced more waste. As we face the next industrial revolution (away from a diet of carbon and fossil fuels) the challenge to us is the opposite and an acute one – we have an abundance of people but dwindling and scarce nature and raw materials. The world's population is expected to grow to 9 billion by 2050 compared to 6 billion today (it was 1.5 billion a mere century ago). Most of the expected increase will be in the developing countries with their thirst for the same energy-intensive gadget-based lifestyle of the developed countries. (Out of the 6 billion today, 2 billion people have no formal access to modern energy – they make do with cow dung, agricultural residue and other solid fuels which are far from healthy. They may embrace the filthiest and most carbon-emitting forms of fossil-fuel energy as soon as they get the chance.) According to the WWF, we would need three planets' worth of resources to sustain this planet if we carry on business as usual based on our existing trajectory of energy and resource use.

Conventional wisdom

The solution and way out of our present and future crisis is not tinkering with more end-of-pipe emission add-ons to our present industrial systems. Conventional wisdom would suggest that we need only to reduce the worst of the negative impacts of our industries to move seamlessly into a future with confidence. Witness, for example, the current thinking that carbon capture and storage used with conventional fossil fuel coal-fired power stations as a key technology-led solution to address our climate change problems.

Such conventional thinking blinds us to the fact that the destructive virtues and inherent inefficiencies of the current industrial world are the result of a fundamental design flaw rooted in the Industrial Revolution and exacerbated by incremental changes. Such fundamental flaws could never be addressed either through regulatory reform or conventional technological progress. Put simply, the crux of the matter is that today's conventional industrial world, powered largely by dirty fossil fuels that emit billions of tonnes of greenhouse gases and toxic materials into the air, land and water is unsustainable in any shape or form for our planet and its citizens. At its worst, it is an unethical way to do business and a thoroughly depressing legacy to leave to future generations of mankind who will inherit our planet from us.

An alternative is captured in the four core principles of WASTEnomics:

- **Waste to zero** involves going back to basics and treating all our input materials, raw or recycled, as valuable resources that we use intelligently so as to virtually eliminate waste

- **Waste to own** reminds us that we own and are responsible for the waste that we generate

- **Waste as societal liability** is the recognition that waste creates long-term liabilities as we use this planet's limited resources

- **Waste as asset** underlines the considerable value which is already embedded in the waste that we discard and which could be upcycled, re-used and recovered.

Price is what the buyer pays. Cost is what society pays.

We know the price of everything but the cost of nothing. Price is what the buyer pays. Cost is what society pays. For example, Americans pay about $1.50 per gallon at the gas pump, but gasoline actually costs up to $7 a gallon when you factor in all the costs. Middle Eastern oil, for instance, costs nearly $100 a barrel: $25 to buy and $75 a barrel for the Pentagon to keep shipping lanes open to tanker traffic. Similarly, a pesticide may be priced at $35 per gallon, but what does it cost society as the pesticide makes its way into wells, rivers, and bloodstreams?

Source: Hawken [1]

Why such a book?

We are on the cusp of a new era in our view and assessment of waste in our societies. To date, significant discussions and writings about waste have for too long been from a compartmentalised, silo perspective. Because of the many different forms of waste, practitioners are mainly concerned with the public responsibility aspects – practical methods of waste collection and disposal which are considered to be part of the public health and sanitation function of municipal and city authorities. Many authors have written on the need to reduce consumption of our planet's resources and to recycle as part of the individual's self help and green agenda. Others are concerned with the technologies needed to treat wastes, from anaerobic digestion to incineration. Others still are concerned with the financing mechanisms needed to put together waste collection and recycling schemes as well as the construction of multi-million dollar waste-to-energy plants.

This book aims to fill the considerable gap between and within these perspectives with a wide-ranging conceptual approach that critically looks at waste from a first principles perspective: Why has the problem of waste arisen? How will our waste grow in the future? How can we address the situation now and in the future? Are concepts of zero waste possible? In other words, how can we bring new concepts, insights, perspectives and ideas to reverse the trend of ever-increasing volume of waste and its contribution to land, water and air pollution on our planet?

The overall theme and question we address is: How can we view waste not as a liability but as an asset? The result is this book, *WASTEnomics*, and its four core principles outlined in chapter one.

We believe this book presents fresh insights into waste. In particular:

- How individuals, businesses and societies are compelled to realise that they 'own' the waste created and generated in their normal daily and routine business activities, and the implications of such ownership of their waste

- How 'waste to zero' concepts could provide fresh and sustainable alternatives to today's conventional, highly polluting, industrial systems

- The creation of liabilities from the 'waste' pollution created directly or indirectly from their activities

- The creation of new industries and players in the waste sector, new business models and the introduction of carbon finance, carbon credits and environmental finance[2].

We paint a different picture of the world of the future with the evidence and examples put together in this book, with many examples of potential success for industries and businesses that successfully understand and appreciate the WASTEnomic logic of new business models, utilising carbon finance and carbon credits and creating energy from waste. Many mature firms are revitalising themselves on the new WASTEnomic paradigm, yet others will be subject to competitive threats never experienced before. New ones are created based on the visionary WASTEnomic principles outlined here.

Caveats

Three caveats to the reader. First, because this is a multi-contributor volume written by experts in their own fields, they bring their particular and specific expertise and angles to the chapters. Secondly because the economics and finance of waste is still a fairly new area of research, there are many gaps in our knowledge and understanding. Many potential concepts and mechanisms are there to be applied from other strands of economics and finance. Thirdly, waste is local – as such, the problems and solutions depend primarily on the demographic and geographical specifics; policy prescriptions that are presented here cannot be applied across the board.

---

2   For an elaboration of the role of carbon credits in the financing of renewable energy projects, see Dr Kenny Tang CFA (2005): *The Finance of Climate Change – A Guide for Governments, Corporations and Investors* (Risk Books).

Structure of the book

This book has four sections:

The first section introduces the problem of waste. Chapter one asks why waste is such a major problem and presents the four core principles of WASTEnomics. In chapter two, Adam Read looks at the issue of ownership of waste and of public engagement.

Section two brings together a series of chapters dealing with the economics and finance of waste. Chapter three, written by leading environmental economists William Hogland and Jan Stenis looks at the economics of waste from the equality principle. Chapter four addresses the economics of carbon finance through landfill gas projects, chapter five assesses the economics of landfill mining while chapter six examines the changing world of waste projects. Chapter seven surveys some critical long-term corporate governance and compensation issues in assessing the financing of low-level radioactive material.

Section three combines a group of chapters under the theme of waste management and waste-to-energy. Chapter eight surveys the disparate waste management practices throughout Asia while chapter nine, by leading development economist Tim Forsyth, assesses how waste-to-energy projects are financed in Asia. Chapter 10 analyses the environmental, energy efficiency and financial considerations of waste-to-energy plants in Europe while chapter 11, by Martin Kurdve, reassesses the waste hierarchy through the perspective and lens of a major corporation like Volvo. Chapter 12, by Adewole Taiwo, analyses waste management practices in Nigeria from a sustainable development point of view.

Section four contains a number of forward-looking chapters dealing with the achievability of zero waste. Chapter 13, by environmental economist Sandra Lebersorger, assesses the possibilities of achieving zero waste through waste prevention, while Peter Beigl, in chapter 14 asks where waste is heading and analyses the future predictive capacity of waste. In chapter 15, Ranjit Singh Baxi looks at paper recycling, recovery and upcycling as a cornerstone of the sustainable use of renewable sources. Stewart Anthony and Jane Fiona Cumming close the book with their chapter: 'Zero waste – the achievable dream' and examine the business and communications aspects of zero waste.

Challenge to developing countries

Developed countries in recent years have adopted higher air emission standards and tougher water quality standards through a mix of public pressure and lobbying from the NGOs with increasing levels of environmental and ecological awareness.

On the other hand, developing countries, especially those in Asia and Africa, are now the production factories and extractive grounds of the world respectively. According to the World Bank, the economic burden on society caused by bad environmental practices amounts to between 2 per cent and 5 per cent of GDP. The WASTEnomics principles are indeed core to the way we manage our manufacturing and extractive capabilities in the future. Through these WASTEnomics principles, we have set the challenge to Asia, Africa and the developing countries to:

- Rise up to the plate to provide basic standards in air, land and water quality to their citizens in a positive and progressive manner through:

- setting tougher air emission and water quality standards in their local factories and in their cities

- enforcing tougher extractive practices, with penalties for the extraction of their (and the planet's) one-time windfall resources

- Negotiate, cajole and demand equal human rights for their citizens from international and global business organisations whose manufacturing facilities – located in these host developing countries – produce lower quality emissions than their own sister facilities in the developed countries or, worse still, consciously or otherwise, dump toxic and other hazardous materials in the developing countries. Citizens of developing countries have the same basic human rights as the citizens in the developed countries.

The regional and global development banks need to provide more support than before to these national governments by adopting tougher financing mechanisms and rules for major projects in ensuring that the tough emission standards of the developed world are not sacrificed in the developing world in the interests of 'development'.

Acknowledgements

We offer our sincere thanks to Celia Cozens, managing editor and John Sivak, managing director at Middlesex University Press, for their support and encouragement throughout this book (and continuing the series of books). Paul Jervis, Marion Locke and Matthew Skipper provided invaluable support.

Our sincere thanks also to Dato' Abdullah Badawi, Prime Minister of Malaysia, John Elkington (founder and Chief Entrepreneur of SustainAbility), Daniel Esty (Hillhouse Professor of Environmental Law and Policy at Yale University), Jonathon Porritt (Co-founder of Forum for the Future), Sir Stuart Rose (CEO of Marks and Spencer) and Sir Richard Branson (chairman of the Virgin Group) for contributing forewords and quotes to this book as well as associated publicity materials. A book venture like this could not proceed without financial support, especially from Tan Sri Dr Francis Yeoh CBE and my previous co-author, Ruth Yeoh of YTL Corporation.

We also thank all the expert chapter contributors who have come together to produce a book such as this. Many thanks for your time, energy, dedication and the commitment to share your experiences in this book[3]. This is the fourth such book that we have produced through the collaboration of expert contributors.

---

3   Responsibility for the contents of and arguments advanced within each individual chapter rests with its authors. We have sought to ensure consistency of style and clarity of expression while avoiding considerable overlap, which is unavoidable in some cases.

Kenny Tang

Thanks to my co-editor Jacob Yeoh who has provided a fresh perspective. His fresh pair of eyes from the younger generation is a joy to behold.

Thanks to everyone connected with Oxbridge Capital, especially Sir Paul Judge (chairman), Gordon Young, Nigel Rich, Peter Pearson, Sir Geoffrey Pattie and Yap Hon Seeng for their continuing support. Grateful thanks go to Peter Kwok, Jeff Paine, Fred Nadler, Paul Smith, Quintin Vello, Sampson Low Lok Peng, Seeto, Joo Teh, See Lip, Patrick Chan, Donny Lim, Vilma Tay, Manshaant Vohrah. Special thanks to Pastor Dr. Peter Masters and Assistant Pastor Chris Buss and everyone at the Metropolitan Tabernacle for providing us with a spiritual home (and Pastor David and Ruth Kay at Whiddon Valley).

On a personal note, special thanks to my patient wife, Lorraine, for her cheerfulness and support and constantly urging me forward to produce books of compelling interest to the world that we live in and for her discussions on the WASTEnomics grid. Also to our three young children: Joseph who is so excited with his own progress in reading and writing as well as designing cars; Hannah, who is beginning to express herself so well and little Timothy who is enjoying himself as the baby of the family (and Helga who took a special interest in looking after our children's reading and writing). They constantly remind me of our considerable obligations to the future generations in terms of our nurture and care of our planet.

I also thank my in-laws, Brian and Christine; Jan and Sylvia; Simon, Rose, Samuel and James, and my extended family of brothers and sisters including Doris, Freddie (and Siew Wah), Lawrence (and Helen), Chelsea, Pat (and Philip), Choon, Sam, Woh, Phing and On; nephews and nieces including Damien (and Daisy), Becky, Kieran, Katrina, Thim Yee, Weng Kern (just beginning his degree course in Australia), Sharon (and Steve), Rena (and Ganesh) (best wishes for their weddings in 2008), Michael (and Pam), Calvin, Peggy, Maggie, Wendy, Yue Wai (Benny), Chia Li, Dr. Weng Heng, Oi Yee, Kah Yee and others.

Special thanks to all at Kennington Lane who have provided such as a fascinating observation post: Kebba Jobe, Ahmad Akhtar, Toya Olufemi, Nathaniel Arko, Mohammad Islam, Qasim Naqvi, Adeshola Gbademosi, Alan Billingshurst, James Byrne, Tsegay Ghebrehiwet, Aderonke Bademosi, Olaitan Okuwobi, Sean Labode, Chuks Onwuka, Edward Boakye, Douglas Owusu, Muhammad Farooq, Kieron Dunkley and Consign 'Announcer' Dhesi.

Jacob Yeoh

This book was made possible by all the wonderful people who have provided assistance through invaluable wisdom, guidance, support and friendship which has made the completion of the book a thoroughly fulfilling experience. I like to thank the Lord for giving us a wonderful opportunity to experience life on what we call Earth, although we must all take responsibility to clean up after us and leave it beautiful and sustainable for future generations.

I thank Dr Kenny Tang for his insightful views and ideas that provoked new concepts and projects that could possibly be implemented. He also spurred me to do some research during my final year at university about sustainable waste projects in the United Kingdom as well as globally in an effort to meet the rising needs of alternative fuel resources, highlighted in the recent UK Energy White Paper.

Finally I thank my father, Tan Sri Dr Francis Yeoh, my grandparents, family and relatives for their encouragement. My father, for his wisdom and new ideas that helped me look at business in a different light as well as my sister, Ruth Yeoh for her encouragement and her ideas from the book *Cut Carbon, Grow Profits*[4] (that she co-edited with Dr Kenny Tang).

This book is dedicated to my mother who taught me humility and strength so that I may apply them to whatever I do today.

Dr Kenny Tang CFA and Jacob Yeoh

WASTEnomics Capital and YTL Corporation

March 2008

References

[1]   Paul Hawken, 'Natural Capitalism'. *Mother Jones*, March/April 1997.

4   Dr Kenny Tang CFA and Ruth Yeoh (2007): *Cut Carbon, Grow Profits – Business Strategies for Managing Climate Change and Sustainability*, Middlesex University Press.

Section 1:
Introduction to the
Global Waste Problem

1

Addressing the global problem of waste – the four core principles of 'WASTEnomics'

Kenny Tang CFA
WASTEnomics Capital

Civilization is being poisoned by its own waste products.

W.R. Inge

Over the last 150 years, since the dawn of the Industrial Revolution, the West, dominated by the USA and Western Europe, has become the richest part of the world. This is the 'positive' and 'good' side of such rising wealth; the 'negative' and 'bad' side of such wealth is that we produce, among other things, more waste than ever before.

These are the by-products of our consumption process, including the goods as well as their packaging (both packaging for sale and the packaging used in the transportation of such goods from the place of manufacture to the warehouse and the point of sale). In addition, enormous quantities of 'waste' are generated in gathering the inputs (such as the extractive process in mining the raw materials) and in the production and assembly processes of these goods and their components. These manufacturing facilities spew tons of chemical waste into the river systems in addition to belching tons of toxic materials into the air. Waste pollution is considered the third major form of pollution alongside air and water pollution.

What are the reasons for the rise in waste?

Rise of consumption and rise in income

We are very good at putting consumption at the heart of our lives. As long ago as 1948, Victor Lebow said: 'Our enormously productive economy demands that we make consumption our way of life, that we convert the buying and use of goods into rituals, that we seek our spiritual satisfaction, our ego satisfaction, in consumption. We need things consumed, burned up, worn out, replaced, and discarded at an ever increasing rate.'

Furthermore, waste generation has positive income elasticity. This means that as income grows, the amount of waste generation grows. We have seen an enormous rise in living standards and income levels in the West followed inevitably by the amount of waste generated. Given the tremendous growth seen in the major developing countries such as China, India and Brazil, the world is braced for an explosion of waste in all forms created not just by the demand for such goods by these fast-growing countries and their consumption appetites but also by the manufacturing facilities built to satisfy the demand from these countries.

'Throw-away' societies and shorter product life cycles

With rising incomes, we have evolved within a short period of history into a throw-away society with more wastefulness than ever before. The globalisation of the world economy means that goods can be made more cheaply and therefore more affordably in most Western societies. For example, white goods such as fridges, freezers, washing machines and modern music systems are present in modern homes. Even though these goods are made to last for many years, they are being replaced at a much faster rate than before, leading to rising mountains of waste in electronic and electrical equipment. Electronic waste is growing three times as fast as other municipal waste. For example, in 2006, the average European threw away 17–20 kg of electrical and electronic products. The amount is rising: a decade ago, the average European was discarding 14 kg of electronic waste. The average British family throws away four such items a year.

The mass production of goods is responsible for the creation of the waste cycle as we know it today on a grand scale. Today's multi-locational production techniques, with import and export of components, assembled goods and materials, encourages greater supply of input materials and thereby greater production waste. Furthermore, such is the demand from our consumption-led society for greater choice that manufacturers produce goods with shorter shelf lives to satisfy the consumers' demand for differentiated products.

Cities, urbanisation and waste

The year 2007 marked a landmark year, as, for the first time in the history of our planet, more people were living in cities or urban areas than in rural areas. Every year there has been a net annual migration of 33 million from the rural to the urban areas, with more than 17 million alone accounted for by China and India.

Cities produce special problems with waste. Due to growing urbanisation, the attendant changes in lifestyle require more waste through convenience foods that use more packaging, such as ready meals and prepared foods; while the increased numbers of shopping and food outlets require more logistical transport as well as packaging of the goods themselves.

Population explosion

Our planet's population has grown from 1 billion more than two centuries ago to 1.5 billion a mere century ago. Today we stand at 6 billion people, and this is expected to rise to 9 billion by 2050.

If we continue with our current patterns of resource use and energy-intensive methods of usage in our homes and production facilities, we will need three planets' worth of resources to enable the whole planet to achieve the same quality and level of lifestyle. Clearly, this is not possible or sustainable. We need to rethink our conventional view of waste, to redefine the nature of waste.

Rise of 'WASTEnomics'?

There are compelling *economic and regulatory* reasons to make waste the key focus of our resource-use strategies and the compelling theme of the next Industrial Revolution. The growing waste explosion is a global multibillion problem crying out for practical and financial solutions.

Regulatory drivers

Driven by Western Europe and the USA, the prominent rise of ecological and environmental awareness that began in the 1960s and 1970s has now reached a tipping point. Such levels of increased environmental interest and awareness among the public at large – driven, on the one hand, by pressure from the policymaking legislators in the European Union (EU), and, on the other hand, by non-governmental organisations (NGOs) such as Greenpeace and Friends of the Earth – have prompted a widespread change in the way we view our obligations to our planet (as in climate change) and to waste in particular. EU legislation has brought about an enlightened and urgent change in the way waste is viewed and dealt with.[1]

Under such a changing, stricter and tougher regulatory structure, waste has gone from being a liability – a waste product to be disposed of – to being seen as a valuable resource from which value can be created.

> *'Waste lies at the heart of every society's environmental challenge. WASTEnomics reconceives of waste as a resource out of place rather than pollution. This fresh perspective offers an important path toward improved environmental results across the world.'*
>
> *Daniel C. Esty, Hillhouse Professor of Environmental Law and Policy, Yale University, and author of Green to Gold*

---

1   It is estimated that in the United Kingdom more than 70 sets of laws and regulations have been implemented since the rapid rise in European Union legislation including recycling levels, emissions standards, etc.

Redefining waste

Waste goes by many names – rubbish, trash, garbage, refuse, scrap or junk. Whatever term is used in different parts of the world, waste is unwanted or undesired material. Waste is anything that is unwanted or unvalued by its owner. Waste is also defined as 'a man-made thing that is, in its given time and place and in its actual structure and state, not useful to its owner, or an output that does not have any owner'.[2]

Because waste of any kind – from the extractive process of raw materials to the production waste of our throw-away culture – is unwanted or undesired, it is not seen as our *own*.

Furthermore, waste is rightly seen as a *liability*. It has a significant cost attached to its collection, transport, clearance, disposal or remediation. This could be in the direct cost of physically collecting and disposing of the waste – our *ownership liability* – or in our *societal liability* whereby costs are incurred in treatment before disposal or remediation of land and water that is directly or indirectly affected in the extraction, production and assembly process. In many less-regulated countries, such waste is disposed of haphazardly – chemical wastes are routinely discharged into rivers thereby causing water pollution, or discharged into the air causing air pollution.

However, if waste is viewed as resource inputs into another process, such waste could potentially be considered an *asset*. If important materials are extracted from the waste stream and reused or recycled, such waste streams are valuable assets in their own right. Given the finite resource base of our planet and the rapid industrialisation in many developing countries, such recovered materials must be considered valuable assets given the considerable energy and costs already expended in the extraction, refining and concentration processes.

Manufacturing could be considered the opposite extreme of a *zero-waste* production process – as much as 90 per cent of all the materials extracted to manufacture ordinary consumer goods end up as waste, while only as little as 10 per cent actually end up in the product itself. Many manufacturers have ensured that their goods have a limited life by built-in obsolescence (as in adding more functionality in newer versions of their products); obsolescence is also ensured by the natural loss of competitiveness due to rapid technological advances or consumers demanding more up-to-date products. The consumer lifetimes of these goods are limited before they are taken off the shelves of retailers or replaced by consumers. For example, such is the rate of consumer change that mobile phones are replaced every 9–18 months compared to a lifetime use of at least 6–10 years.

Turning waste liabilities into assets

The new economics of waste is driving a historic change in how individuals, companies, businesses, governments and societies view waste. In other words, the new world

---

2   See Adam Read in Chapter 2 of this book.

of WASTEnomics sees waste not as a liability incurring a cost in its removal and safe disposal, but as an asset for which a market value exists.

> *'The four principles of WASTEnomics presented, explained and promoted in this book certainly draw on best practice – but they also move well beyond, to signal some of the new directions that the twenty-first century economy will take us in.'*
>
> *John Elkington, Chief Entrepreneur and Founder of SustainAbility*

The core principles of WASTEnomics: ZOSA

The new era of WASTEnomics is governed by four powerful core principles that are replacing some of the old tenets of waste:

- waste to **Z**ero
- waste to **O**wn
- waste as **S**ocietal liabilities
- waste as **A**ssets.

Z: *waste to Zero*

Waste to zero is the core, overriding principle. We have to go back to basics and treat all our input, such as raw materials that we take from this planet, as valuable resources that we use intelligently and that could be returned to the soil. It is especially important to ensure equity with due regard to the needs and considerations of future generations.

Waste to zero is not an alternative disposal technology, in direct competition with landfill and incineration. It is a principle of action for change and a diverse, flexible range of policies, technologies and actions designed to ensure that we utilise our resources and materials efficiently and effectively. It requires a whole-system approach to revolutionising and redesigning the way resources and materials flow through society. Waste, under the old paradigm, was seen as a by-product and a signal of design failure; therefore, we need to go back to basics from a planetary point of view.

Would it be possible to design the goods of tomorrow so simply yet so intelligently with a view to the future that, when used, they have a positive and restorative, ecological and environmental footprint, in contrast to the goods of today that have a negative and destructive, ecological and environmental footprint? Would it be possible to produce goods that use safe and positively wholesome, natural materials that can either be returned to the earth to replenish the land or recovered by their producers to be reused? Or machinery and equipment designed for disassembly and reassembly?

Can we design long-life infrastructure assets such as buildings and communities that exist in close proximity with our nature – and use energy from renewable sources, such as the sun, tides, waves and wind; recycling energy throughout the building or community; storing wind and solar energy; purifying water and harvesting rainwater? We must use materials that breathe easily and encourage energy conservation. In other words, we need not just a zero-energy building and community but a zero-waste building and community, too.

The model of the living city that we can provide for future generations is one that allows people and the natural world to coexist and thrive together positively – one supporting the other. Designing, planning and constructing the natural cycle infrastructure to support these living cities of the future must be one of the defining challenges for the global citizens of the twenty-first century and beyond.

By this principle, we should design goods that virtually eliminate waste or have zero waste intrinsically – not just during the production, assembly and distribution process, but also during the consumption process.

Material flow

How can we critically look at the complete material flow management in a natural and positive way – not just waste management at the end of the food chain? We have to think of 'earth-to-earth' rather than 'earth-to-waste' whereby we dispose of resources that we have expensively gathered and refined into a hole in the ground for many years, if not decades. How can we ensure that the materials we use are properly recycled with appreciation in value and purpose? Today, goods and materials are downcycled in value and in purpose – and eventually after another bout of (downward) recycling, they are still on a one-way, 'earth-to-waste' trip to a hole in the ground or to the incinerator.

Production

Can we redesign production systems to minimise use of raw material inputs or redesign production systems to make use of recycled materials? At worst, we need to recycle – everything from the unused factory input, the leftovers from production or the unwanted factory outputs.

Consumption

The need to ensure responsible consumption, whether in the office, factory or business, is obvious. Packaging is a by-product of our modern consumption society – from packaging for sale (e.g. wrapping and packaging of raw foods as well as ready-made prepared foods) to packing for retail use (e.g. plastic bags) to packing and packaging for transport of goods (e.g. cardboard boxes, etc.). How can we design goods that require minimal packaging or use packaging that dissolves organically and naturally?

Creating industries of the future

This core WASTEnomic principle could spark many natural but creative and out-of-the-box ideas, concepts and ways in which these radical changes, together with our reconnection to the natural life of the planet, could revitalize and re-energise our pollution-led industries, our pollution-filled cities, our pollution-financed economies and our growth-led nations. In turn, these principles could transform the way we make and consume goods and thereby transform today's global citizens and their responsibilities into accountability for the use of the earth's resources to our planet and future generations.

Thus, waste to zero is a driver for creating sustainable communities and presents immense new opportunities for employment and local economic development.

O: waste to Own

Waste – once produced and generated by us, as extractors, producers and consumers – must be treated as our own. It is therefore our responsibility to society and to the planet to discard such waste (especially that containing embedded toxic materials and metals that are hazardous, harmful and lingering) in a responsible and productive way.

A well-known consumer waste that we regularly discard is plastic; each individual is said to discard some 65 lbs of plastic into landfills each year, while less than 5 per cent of plastic is recycled. Worse still, storms flush plastics downstream and ultimately into the oceans, which are especially susceptible to plastic pollution. It takes longer for the sun to break down plastic in the ocean than on land because of the ocean's cooling capacity.

Most plastic floats near the sea surface, where some is mistaken for food by birds and fish. Plastic is carried by currents and can circulate continually in the open sea. Plastic pollution negatively affects trillions upon trillions of ocean inhabitants and ultimately humans.

A particularly galling discovery recently was the 'plastic soup' of waste in the Pacific Ocean that now covers an area the size of the continental USA.[3] Plastic in the ocean is one of the most alarming of today's environmental stories. Plastic is forever: because plastic does not biodegrade, no naturally occurring organisms can break the polymers down. Instead, plastic goes through a process called photodegradation, whereby sunlight breaks it down into smaller and smaller pieces until there is only plastic dust. But plastic remains a polymer. When plastic debris enters the sea, it can remain for centuries, causing untold havoc in ecosystems.[4]

Producers of goods must internalise the costs of disposal of such 'waste' from the extraction as well as the production process, including extended producer responsibilities. Individuals and consumers must internalise the costs of disposal of both medium-life and long-life electronic and electrical consumables that they purchase, perhaps through an additional levy at the point of purchase to pay for its eventual disposal. The onus is to fund sites where consumers can discard electronic goods to be properly disposed of. (Medium-life goods include mobile phones, printers, computers, monitors, laptops, MP3 players, etc., while long-life goods include cars, washing machines, refrigerators, etc.)

3   'The world's rubbish dump: a garbage tip that stretches from Hawaii to Japan',
      Independent, 5 February 2008
4   Material for this section on plastics came from Algalita Marine Research Foundation:
      http://www.algalita.org

Many electronic and electrical products discarded by consumers contain mercury, cadmium and toxic flame retardants. Cathode ray tube televisions and computer monitors are a particular problem because of the amount of lead in them.

Government

Governments should promote wider policies that encourage the safe disposal of electronic goods (e-waste) such as the EU's Waste Directive on Electronic and Electrical Equipment (WEEE), including stiff penalties for non-compliance.

Finance sector

The finance sector should devise new mechanisms to help the business sector recognise its long-term liabilities, including extended producer responsibilities. Can we learn from the financing of very long-life nuclear power stations and the need to build in the 'financing' of the end-of-life costs of the decommissioning process? How can long-life assets (such as major power stations) include the end-of-life costs of decommissioning into the initial project financing mechanism?

Business

Businesses should redesign goods so that they can be disassembled and the parts reused, and should limit the use of toxic materials in their goods. They should promote marketing campaigns that encourage their customers to return their products to be reused in other countries or to have their precious materials recovered.

Box 1. E-waste: in need of a cure

China needs an e-waste solution badly. Most of the world's estimated 40 million annual tonnes of e-waste – production scrap and discarded appliances – ends up illegally dumped in China. Domestic e-waste is also rising rapidly. The Chinese already trash 40 million personal computers every year, according to the UN.

'China's Massive High-Tech Waste Woes', *Business Week*, 9 August 2007

S: *waste as Societal liabilities*

We are the creators of our own waste – be it from the extractive process of obtaining raw materials, the production process of manufactured goods using components and modules including discarded production wastes, or from the consumption process. We extract raw materials from our planet such as coal, tin, iron ore, etc.

What are the visible signs of such extraction? Polluted iron-ore lakes with metal residues in the water and vast tracts of disused land; mercury released into the air by coal extraction; polluted soil, from the thick masses of black sludge left in lakes after palm oil (used in a wide variety of industries) has been extracted from specially planted palm trees; vast quantities of chemicals belched by our factories into the air.

As extractors and exploiters, and also as consumers of resources, we must be responsible for, and own up to, the direct and indirect damage that we cause to the ecology and environment of the planet. These are the negative externalities or *societal liabilities* that we create. Among others, this includes the wastes that we extract and discard from the raw materials, the 'damage' to the exploited and mostly barren land and polluted lakes, and the wastes from the production process as well as the considerable quantities of chemicals and toxic materials that inevitably find their way into the air, land and water systems. Such 'wastes' from our extractive and production processes pollute the air, water and land, with considerable damage to air and water quality that requires significant investments to put right.

Ironically, even those that recycle our goods and recover the metals and materials end up creating huge societal liabilities through backyard recycling techniques that release pollutants into the air and land.

Box 2. Backyard recycling that does more harm than good

A 'backyard recycling' industry has grown in Guiyu village in China. Some 80 per cent of the 132,000 villagers are now involved in e-waste. They extract metals like gold and copper, and repair components to sell to factories. But lack of expertise means that valuable materials are crudely extracted, often releasing pollutants in the process.

This lack of control is exemplified by the illegal e-waste trade. Although China has ratified international laws and implemented domestic regulations banning e-waste imports, conditions in villages like Guiyu have only worsened.

'China's Massive High-Tech Waste Woes', *Business Week*, 9 August 2007

Governments

Governments should adopt tougher regulations on air emissions and water standards, covering the need to monitor air and water quality regularly. After all, governments that neglect such basic health measures and stick their heads in the sand are in fact storing up problems for tomorrow. Citizens who work and live in these affected areas will face greater health risks and therefore higher health costs in the long run. This has great implications for the productive and human capacity of any nation.

Finance sector

Financiers and investors should exercise greater care in financing the extractive industries and manufacturing processes that damage air, land and water quality. Extensive environmental impact assessment should be carried out. Financiers should design built-in financing mechanisms that recognise the polluters' responsibility for the ownership of these wastes and the societal liabilities that they create.

Businesses

Businesses need to be more responsible and adopt common standards across their entire group of factories. It is very encouraging to see that car engines being built in China are subjected to tougher emissions standards than in the USA. Very often, the reverse is true. While we see tougher emissions standards in the USA and EU, similar factories in developing countries play hard and fast with the local rules and adopt lower 'local' standards.

A: waste as *A*ssets

All wastes – whether that from the raw materials extraction process, residues from the production process, or wastes left over from our consumption – should be viewed as valuable materials. These recycled materials, reused materials, and recovered materials all have great economic value in a resource-constrained world, especially in a world of rising energy as well commodity prices. Much energy and great resources have been expended in extracting and producing metals, and it is to our significant long-term cost that much of these 'wastes' are currently left in landfills or discarded in mines and lakes.

Much municipal solid waste is now separated for recycling while the remainder is incinerated in waste-to-energy plants due to the higher cost of landfills. The new business models include charging a gate fee for processing the waste, obtaining a price for the recovered materials, and selling the power generated from the incineration plants.

Government

Governments should actively promote policies and investments that enable safe and proper recycling, recovery and reuse of materials, including plastic, paper, metal, glass and wood.

Business

Businesses should see high-quality recycling, upcycling, recovery and reuse of metals and materials as major opportunities in the WASTEnomics era.

Finance sector

The finance sector should evaluate and invest in new sunrise industries of the future in waste management, recycling and upcycling facilities, waste-to-energy plants, metal- and plastic-recycling businesses, etc. These businesses are thriving on the back of a number of key business drivers, such as rising standards of air and water emission regulations by many local, city and national authorities; and rising energy prices caused by rising oil prices, need for energy security, adoption of carbon pricing and the like.

Applying the principles

In applying the WASTEnomics principles to ourselves and our organisations, we need to consider our starting point (see the WASTEnomics grid in Figure 1). In other words, which are the major visible points of our waste footprints and in which quadrant are the majority of our waste footprints located?

Owner (liability) quadrant

Most, if not all, individual organisations start here (the bottom left corner of the road map – Figure 1). It is no surprise that this is the starting point of our waste journey through the road map. This is the waste stream from our households and from our business operations – be it household waste, production materials discarded or unused appliances. It is therefore critical that we take ownership of such wastes and be responsible for their proper disposal.

Major business opportunities are found in this quadrant from optimising production to reduce waste, reuse of production waste, etc. Increasing opportunities are also found in the collection of e-waste.

Societal liability quadrant

This is the quadrant that recognises the negative externalities or our societal liabilities to the planet in terms of the damage to the air, land and water from our direct operations in extracting and using the resources of the planet. Unsurprisingly, very few organisations admit to such liabilities, and even fewer organisations record these liabilities on their financial balance sheet.

Major business opportunities are to be found in this quadrant from identifying and quantifying such liabilities on the company's balance sheet, and the devising of financial mechanisms that enable the long-term liabilities to be adequately funded. The ratification of the Kyoto Protocol, which limits the amount of the six greenhouse gases emitted into the atmosphere by major developed countries and businesses, has spawned a thriving business in the area of carbon finance and carbon trading. This area is set to grow with new opportunities in, among other things, water rights.

Asset quadrant

This is the quadrant that has seen the most significant volume of business activities as well as investment ranging from high-quality recycling of metals and materials to the construction of multimillion-dollar investments in waste-to-energy plants. Municipal and city authorities are recognising the need to build adequate kerbside collection and recycling facilities given the current unsustainable methods of disposal through landfills.

Waste-to-zero quadrant

This is the quadrant with the most potential for the future, not just for saving the planet but also in terms of the creativity needed to address the problem. We expect to see major and significant investment in research and development as well as in strategic business opportunities that are likely to appear in this quadrant.

SOCIETAL LIABILITY	**Z**ERO
Waste-as-societal liability (e.g. damage to land, air, water quality)	Waste-to-zero (core principle, that is, a whole-systems approach to usage of resources and materials)
OWNER LIABILITY	**A**SSET
Waste-to-own (owner liability for e-waste or electronic waste, production waste, household waste, etc.)	Waste-as-asset (e.g. recovery of metals, high-quality recycling, waste-to-energy plants, etc.)

Figure 1. The WASTEnomics grid

Concluding remarks

The new era of WASTEnomics will drive a seismic change in how individuals, businesses, financiers and governments view waste now and in the future. The core principles outlined here will revolutionise the way individuals, businesses and societies view their ownership of and responsibility for waste; and their use of resources in our societies; and the roles of public engagement. These principles will govern our innovation plans for how we design almost everything in our society in the future. They will drive governments, including city and local authorities, to adopt tougher air and water standards for their citizens. They will open our financing mechanisms to more responsible and sustainable businesses. They will compel the forward-thinking way in which businesses design and produce goods in the future.

WASTEnomics is a new framework for the individual, business and government to look at the issue of waste and the use of our planet's resources. If the growing problem of waste from our homes and our factories, in our cities and in our societies, is to be successfully tackled through responsible use of our planet's resources and not get out of hand, the four core principles of WASTEnomics should guide every decision made by individuals, businesses, financial institutions and governments today and in the future, in every boardroom, every design studio, every council meeting, every financial institution, every product innovation meeting, every classroom and university lecture theatre, and every marketing campaign. That is the scale of our challenge!

Box 3. WASTEnomics – 11 steps to revolutionise waste[5]

Whole-systems approach to waste

Act (on your own) waste

Societal waste

Towards zero waste

Educate waste

Necessary (waste)?

Own waste

Maximise waste (as assets)

Invent waste (ways to upcycle?)

Costly waste

Strategic waste

5   This section benefited from several conversations and discussions with Lorraine Tang.

2

The social agenda in waste management – from individual ownership of waste to public engagement

Adam Read
Hyder Consulting

This chapter discusses a key dimension of 'WASTEnomics' – that of addressing the liabilities of the waste that we generate through redefining ownership of our waste and an increased role for public engagement.

Introduction

Waste and its management can trace its roots back to the earliest stages of civilisation, ever since man stopped being a hunter-gatherer and began to form permanent settlements, there has been a build-up of waste products that have required collection and disposal in one form or another. The edge-of-town 'dumps' of ancient Rome and Troy, the burning of wastes by the Maya and the Inca, or the *'rakers'* of medieval London collecting wastes from the streets are classic examples of early waste-management systems and infrastructure.

With increasing industrialisation, population growth and levels of affluence, the waste products of our societies have become more visible and more of a problem. However, it is easy to forget that 'waste' in its modern form was not truly a worry until the Industrial Revolution in the late eighteenth century, which enabled mass production techniques and greater import and export of goods and materials. This encouraged greater choice, shorter product lifetimes, and a plethora of new composite materials and products.

We have evolved into a throwaway society with more wastefulness than ever before and with more synthetic materials that do not naturally degrade (plastics, synthetic fibres, etc.). Today, in the global economy, goods can be made more cheaply and so, many of us can afford to be wasteful as never before, whether it is because something is 'out of fashion' or because we want a new model or a different colour, or because the item is designed to last for a shorter time, or we do not look after the item as well as we used to. Whatever the reason, materials and goods are becoming waste quicker and more frequently than ever before. Rising mountains of waste have now become a major concern, and at the heart of the issue are global production and consumption patterns.

On top of this wastefulness inherent in many modern societies is the parallel problem of population growth and the pressures that this brings in terms of resource use and waste production. Countries experiencing the most pronounced population growth are often the transition and developing economies of Central America, Africa and Asia – countries whose populations are not only growing rapidly but are also aspiring to Western standards of living and consumerism.

These three related issues of population growth, consumerism and affluence are leading to a crisis in waste production and management for many places around the globe, and this is a problem that will only get worse in the short term. How can we in the West expect African and Asian peoples not to consume and enjoy freedom of choice in their goods and materials when we have been doing so for the last 50 years? All we can do is help them understand the problems and assist them in putting infrastructure and policies in place to control the problem in the short term.

There is little doubt that the general public (or society at large) and waste management go hand in hand. If it was not for consumers, we would have little waste to concern ourselves with; there would be less choice in products, less need for a new model, and less demand. The evolution of the consumer society since the 1950s (post-World War II) has seen a proliferation of new waste-management techniques and approaches, mirroring the more visible nature of waste in our societies today and the need for its collection, treatment and ultimate disposal.

The development of products with shorter lifetimes (built-in obsolescence) and the growth of consumer power, marketing and globalisation were addressed in a seminal book by Vance Packard, *The Waste Makers* (1961). According to Packard, 'A society in which consumption has to be artificially stimulated in order to keep production going is a society founded on "trash and waste", and such a society is a house built upon sand!' Wise words indeed and very prophetic given the marketing-led world we live in today.

Society, according to Packard, was being urged to consume more and more for fear of the magnificent economic machine turning on them and devouring them! The average citizen of the USA consumed twice as much in 1960 as in 1940, while 40 per cent of the things he or she owned were not essential to physical well-being, but were optional or luxury.

Furthermore, Packard believed that we had begun to follow a path that can lead only to pain and failure. He believed that the fundamentals of consumer life were undermining the environment and our society's sustainability. He was a leading advocate of more sensible production and consumption, realising the unsavoury impacts of waste on our environment.

We have come a long way since 1961, and although we are now more sophisticated, and have better knowledge and understanding of our environment and our impact upon it, we are still a market-driven global economy focused on 'the new, the shiny and the purchase', and not on the implications of these decisions on waste, and its collection, treatment and disposal. We have yet to see the real impact of consumption on the transition- and developing-economy nations, and the prophecies of Vance Packard may still come true for the rest of world, and for an ever-expanding population. I fear that things will get worse before they get better.

Definitions

People and waste are related by definition. According to a European Union (EU) directive, 'Waste is any substance or object, which the holder disposes of, or is to dispose of pursuant to the provisions of national law in force.' [1]

It would appear that 'waste' is simply a thing that people do not want. The definition does not, however, indicate why the object became unwanted, and consequently, we cannot tell how objects could be wanted again, returning the items to non-waste status.

The categorisation of waste by Eva Pongracz is as follows[2]:

- *Non-wanted objects that were created either as not intended, or not avoided, and have not been assigned a purpose.* Into this group belong outputs with negative market value, non-useful by-products, emissions, processing and process wastes, cleansing wastes, etc. They are created by people but they have no purpose.

- *Objects that were given a purpose with a finite function, and thus were destined to become useless after fulfilling that functional specification.* This group includes single-use products, most packaging, single-use cameras, disposable nappies, etc.

- *Objects with a well-defined purpose, but whose performance ceased to be acceptable.* This is the most typical waste group: obsolete, broken or spoiled products, non-rechargeable batteries, demolition wastes, etc. The loss of performance may be due to a fault in structure or state.

- *Objects with a well-defined purpose, and acceptable performance, but whose owners failed to use them for the intended purpose.* In this category belong

---

1   EC Directive 75/442/EEC (1994).

2   Eva Pongracz in her thesis, *Redefining the Concepts of Waste and Waste Management* (2002).

products used to excess, products that go beyond their target (e.g. artificial fertilisers that are washed out from the soil), and products perfectly functional, but which the owner disposed of, simply because he did not want them any longer.

The interesting thing is that people assign purpose and evaluate performance; therefore, 'waste' is simply a manifestation of society. The last category shows just how important the role of people is in recognising and producing wastes. A perfectly functional object can be labelled as waste, just because one person finds it not useful, or fails to use it, while the same thing could be useful to someone else, or at some other time or place. This intriguing waste class has triggered the need to consider the concept of ownership.

Redefining ownership of waste

Waste should be redefined as *'a man-made thing that is, in its given time and place and in its actual structure and state, not useful to its owner, or an output that does not have any owner'*. This means that the same thing may be waste or non-waste in different times, places, or for different people, and suggests that waste is a dynamic object. Logically, *the role of waste management is to find a new ownership and/or give a new purpose to the waste,* and the wider role of waste management rather than blindly continuing to focus on logistical and technological issues of collection, treatment and disposal.

Focusing more attention on the issue of ownership would undoubtedly help to develop the reuse and recycling agendas, and would support waste-prevention ideals.

Most would agree that the role and importance of individual action in waste management is significant. However, the nature of the human–waste relationship depends greatly on awareness. When it is about consumer wastes – the focus of this chapter – no legislation can be as effective as a well-informed, environmentally conscious, ethical public. The more people are aware of the ethics, the implications and their impacts, the better the quality of this link, and thus the stronger their commitment.

Research has shown from one society to the next that awareness of waste and its management (knowledge of causalities) is generally rather low. For ordinary individuals, current municipal waste-management schemes make the transfer of responsibility for the 'unwanted objects' very simple. People are not even aware of the fact that when they discard unwanted objects, they are actually giving up their ownership of them.

Most people would not question the morality of this 'transfer' nor do they generally care about the fate of the discarded object. Apart from when waste collectors go on strike, or when there is a proposal for a new landfill site or incinerator near their home, few even appreciate the services of waste collection, treatment and disposal. By embracing the ownership concept, we could help to change this. If everyone became conscious of being owners, and were aware of the responsibilities associated with ownership, people would be more careful of giving up their ownership, and would consequently be more careful in choosing a product or service.

Surely this would be the right path for us to take, given the ramifications of continuing with our current approach and the number of consumers worldwide now contributing to global waste production.

The public's role in waste management

So what roles do the public play in waste management?

- We have already mentioned that they are waste producers and should be made more aware of their intrinsic role in creating the waste mountains facing us today. The public are also service users of waste-collection systems, and therefore need to understand how these services work, what they cost, and what the ramifications are if they are stopped.

- Essentially, the public are the funders of municipal services (collection, treatment and disposal) through user fees, rates, charges or taxes. As such, they have a democratic right to ask questions and voice their concerns, and they can be a significant political force.

- Just as important, and increasingly more significant with every day, the public are the users of segregated dry recycling and organic waste collections, which in the West are an essential element of meeting EU directive targets.

- The public also have a vital role to play in ensuring that new facilities are built, participating in planning inquiries, and presenting evidence to block certain technologies, sites or proposals.

These roles and the interaction of the public[3] with providers of waste-management services will be discussed in the next section, where public engagement, participatory planning and awareness raising are discussed in more detail.

Clearly, the public are involved at every step of the waste-management chain, even if they do not recognise it. Waste managers must ensure that they recognise the public's involvement at every step, encouraging them to take ownership for the wastes they produce and ensuring that they appreciate the concept of producer responsibility more fully.

---

3   More specific to transition and developing economies are the scavengers – the individuals whose livelihood relies on scavenging from the waste that society produces. These groups have a significant stake in 'traditional' waste management systems in many countries, and as such must be a key stakeholder in any proposed change, development, or proposed improvement. We simply cannot impose a new waste-collection system with engineered landfill sites and ignore the needs of these scavengers. We must understand their needs and consider how their welfare can be addressed and improved, whether through concessions to scavenge for recyclables at the landfill site or as recycling collectors in the cities where they live.

We can no longer afford for waste to be 'out of site, out of mind'. People are essential to the success of many new waste-management systems, and need to take responsibility for the wastes they produce. There is recognition among waste professionals of the significance of the general public in waste management, and we now fully appreciate the public's role, and are reacting to this. This is something that waste managers around the world are currently tackling, attempting to communicate with the public and engage them more actively in solutions.

There is undoubtedly a need for a major change in public awareness and engagement relating to waste and its management. Underpinning the drive for greater engagement, consultation and participation worldwide has been the Aarhus Convention. The UNECE Convention on Access to Information, Public Participation in Decision-making and Access to Justice in Environmental Matters was adopted on 25 June 1998, and requires a new kind of environmental agreement, linking environmental rights and human rights.

The Aarhus Convention grants the public rights and imposes on parties and public authorities obligations regarding access to information and public participation and access to justice. This is perhaps the key piece of legislation now focusing the minds of local government. When local governments devise new waste strategies, specify contracts for services, or plan to build new infrastructure, they will be required to consult with all stakeholders as never before. This is a challenge that I am sure every authority is tackling head on, but one for which most authorities may not be ready for, whether in the UK, elsewhere in the EU or, more importantly, in a transition-economy country.

The UK government's policy statement on more sustainable waste management clearly states that 'local authorities are expected to consult with their communities both on their waste management strategies and plans and on any new schemes in order to ensure that local communities understand the need to manage the waste they produce; have a say in deciding the best solutions for managing it; (and) gain economic value from recovering valuable materials from their waste' (Waste Strategy 2000).

Clearly, there is now more attention than ever before on waste management. Whether it be driven by international conventions requiring greater public-engagement practices that ultimately require greater public understanding and involvement in decision-making, or by the growing mountains of waste being produced, or because of EU policies requiring a major change in waste-management practices and public participation in the services on offer.

The need for public consultation and engagement

It is now an unwritten rule that any kind of local authority planning, strategy and service development process must be underpinned by participatory decision-making processes that involve a wide range of stakeholders.

My recent experience has been in a number of waste-management master plans for countries, regions and cities in the developing world (funded by international donor agencies) – all of which have involved extensive stakeholder reviews, participatory planning protocols and widespread consultation. In the last few years, participatory planning[4] and awareness raising of waste management has taken on a new lease of life in the UK, as politicians make difficult decisions about future waste-management services and facilities and have asked the general public for their opinions, involvement and support.

There is an increasing recognition in industrialised countries, and to some degree in transition economies, that all of the 'stakeholders'[5] in waste management should be involved in, and informed about, decisions which might have an effect on their lives, in order to:

- increase their trust and involve them in the decision-making process;
- provide them with accurate information, and overcome some of the misinformation that they may have received about specific technologies or sites;
- get their support for national waste strategies and local delivery of services;
- get local acceptance of the need for facilities required to meet the policies;
- comply with the Aarhus Convention, which 'requires early public participation before a decision is made'.

Table 1. Involving the public

Information giving	Consultation	Consensus building
Leaflets	Surveys	Community advisory groups
Newspaper/TV advertising	Questionnaires	Citizen juries
Exhibitions	Public meetings	Planning cells
Guides/books	Telephone helpline	Workshops

A continuous public information programme can provide background information on wastes, their reuse and recycling, and their treatment and disposal. This can be done through the local authority website, posters and leaflets, or displays at council offices.

This can help to set the management of wastes within its wider environmental context, it can demonstrate ways in which the public can reduce its own impacts,

---

4   For more information on participatory planning methodologies, see "The Strategic Planning Guide for Integrated Solid Waste Management in Developing Economies", produced by ERM (2001) on behalf of the World Bank.

5   Stakeholders are described as 'all of those who have rights, responsibilities and interests in the issues', namely the general public and other waste producers and service users.

and it can reinforce the message that waste management is itself a public health and environmental protection measure. Often such an programme can be costly and may not be high priority as authorities seek to make rapid changes to ensure they satisfy statutory commitments from their national or international authorities.

There is an important distinction between providing the public with information in an attempt to educate (a one-way process, as noted above), and real public involvement through consultation and dialogue. Consultation can be more expensive in terms of funds and staff time, and often requires specialist skills of facilitators, consultation experts and consultants. Consultation is now widespread in industrialised countries, but is often seen as a luxury in developing economies and is thus not as commonly practised. However, international donors funding large-scale waste-management strategy and service-delivery programmes in these locations usually insist on consultation and participatory planning approaches.

Whether authorities adopt consultation or consensus-building approaches, the success of the engagement process will come down to whether the public effectively participate. The chances of the programme succeeding can be enhanced by:

- ensuring that the process and participants are representative of all key stakeholders;

- giving individuals a chance to 'input to the agenda', and building their trust and ownership of the waste problem;

- allowing sufficient opportunity and time for 'deliberation' through dialogue and debate, direct questioning of experts, panel debates and questioning, and, where appropriate, engagement with 'dissent' and those that disagree, in a search for common ground;

- allowing the public to challenge experts and giving them time to check claims and build their own knowledge and understanding of the issues;

- ensuring that the participants' contribution is acted upon and that their involvement improves the process;

- making the process transparent and open, so nobody can criticise the final result.

Public awareness raising in the UK

The 1990s was a decade of slow improvement in recycling across the UK. Authorities continued to develop recycling schemes in order to meet the aspirational municipal solid waste (MSW) target of 25 per cent recycling by 2000. Most authorities never got close, because they did not have to.

This all changed in 1999, and almost overnight authorities began to act. Suddenly, 'high-diversion' recycling became the talk of the town and authorities began to invest, experiment, and test new recycling services. In only 12 months, Daventry District Council had increased its recycling rate from 9 to 40 per cent, through an integrated recyclables and organic collection scheme and by restricting the residual waste collections. But why did all this change suddenly begin?

We must turn our attention to Brussels to understand the current UK situation. The waste-management policies of the EU are increasingly being driven by the principles of sustainable resource management, resource efficiency, extended producer responsibility, climate change and global warming.

The EU landfill directive (1999) has set the following challenging targets to limit the amount of biodegradable municipal waste (BMW) going to landfill:

- by 2010, no more than 75 per cent of total BMW produced in 1995 to be landfilled;

- by 2013, no more than 50 per cent of total BMW produced in 1995 to be landfilled;

- by 2020, no more than 35 per cent of total BMW produced in 1995 to be landfilled.

In the UK, in response to the landfill directive, waste strategies have been produced for England and Wales (2000), Northern Ireland (2000), Scotland (2003) and Wales (2002). For England, the government, in its Waste Strategy 2000, has set a household waste-recycling/composting target of 33 per cent, and a municipal waste-recovery target of 67 per cent, both to be achieved by 2015. Intermediate recycling and recovery targets have been set for 2005 (25 and 40 per cent, respectively) and 2010 (30 and 45 per cent, respectively).

Through government funding there has been a rapid expansion of kerbside recycling infrastructure, with almost every household now offered some form of kerbside recycling. However, no matter how rapid the expansion of kerbside recycling schemes, there is a limit to what they can achieve, as exemplified below:

- recyclable/compostable proportion of the waste stream: 70 per cent

- kerbside households served: 90 per cent

- households participating: 75 per cent

- effectiveness of participants: 90 per cent

- waste collected for recycling/composting: 42 per cent.

And this is why the public are so important in terms of waste management and recycling in industrialised countries where the focus is on targets, service efficiencies and league tables.

If UK authorities want to exceed 50 per cent diversion, they must improve household participation (to 95 per cent) and their service's effectiveness (to 95 per cent). This will not be easy, and thus there is a real need for innovative and successful awareness raising and public engagement campaigns to ensure that residents are aware of the services and fully participate in them. It is about empowering them and giving them ownership.

Clearly, it is time in the UK to go beyond consultation and stakeholder engagement and look closely at residential awareness, understanding and participation in schemes once they are implemented. We have strategies and services in place, but we are still failing to meet our targets, one of the most common barriers being the awareness, attitude and participation of the public.

We are entering the age of public ownership of waste management, and without the public we will not succeed in achieving more sustainable waste-management practices and meet tight legislative targets in the UK.

Box 1. UK campaigns

Campaigns such as Rethink Rubbish, Doing Your Bit, and the new WRAP National Campaign can help to move the public towards recycling, but more localised campaigns and programmes must be developed to build upon these foundations and encourage residents to act by participating effectively in the recycling schemes provided. Campaigns such as the Don't Let Devon Go to Waste, the Essex War on Waste, and the Western Riverside Rethink Programme must be seen as examples of best practice, and other local campaigns developed to replicate their successes.

More and more authorities are now looking to use best practice in public engagement, education, awareness and participation to ensure they meet their targets, and guidance is available on stakeholder engagement (see the Environment Council's guidance note) and on doorstep promotions campaigns (downloadable from the GLA website) to assist authorities.

If we do not get full participation and effective recycling services, the diversion levels necessary to meet EU standards will require significant investment in mechanical and biological treatment (MBT) and waste-to-energy facilities, neither of which will prove to be particularly popular with the public in the longer term because of the size of facility needed. Nor will they resolve the issue of the public's ownership of waste, and they encourage the 'out of site, out of mind' perspective discussed earlier. In many respects, it is the public's choice, and we must engage them in the decision-making process and ensure that they know what the consequences of failure will be.

Authorities in the UK are now at a crossroads and need new waste infrastructure; so by engaging the public, they are encouraging their wholehearted participation in recycling and ensuring they know what is at stake in terms of new facilities and services.

However, not all countries have the same agenda for their waste-management public awareness raising as in the UK, because they are at different evolutionary position in terms of their waste-management policy and services.

International awareness issues

The UK public are different from many of their EU counterparts in that the UK public do not respond so well to 'direction', and therefore more voluntary approaches have had to be used to induce participation in recycling schemes, and this has required innovation and funding of awareness campaigns.

For example, the Scandinavian countries have a more engaged and participative public, and so their campaigns have tended to focus on specific materials and products and on changes in service delivery.

However, in EU Accession Countries with transition and developing economies, the focus is usually on waste collection-related issues and ensuring that people use and pay for the services on offer, a far cry from the recycling bias currently seen in the UK and other Western countries.

The drivers for waste-management improvement and subsequent awareness raising also differ from one country to another. In the developing world, the focus is very firmly fixed on community health and safety and external agency requirements.

Box 2. Awareness raising in Egypt, Mali and Russia

In both Egypt and Mali, a great deal of time was spent on educating local communities about the heath problems associated with throwing rubbish into the irrigation canals, the problems of stagnant water, insect-spread disease, and the poor health of their children. This type of campaign is quite powerful and did help to change local attitudes and practice.

In Volgograd (Russia), the emphasis was on stopping the home burning of waste, which, given the changing nature of the waste stream (higher proportion of plastics today than ever before), was causing significant respiratory problems for the older residents. Again a targeted campaign highlighting the link between bad waste management and poor health was initiated.

In many respects, the transition- and developing-economy countries are experiencing the problems that were faced in the UK and other industrialised countries in the early nineteenth century. The public health impacts of waste accumulation in the streets and the problems of cholera and rats led to the development of public health legislation and more effective and regular waste collection and disposal services.

The need for public engagement and education is more vital in countries where waste is still a health threat than in the UK, where the focus is on recycling, but some transition-economy countries have already started to address recycling infrastructure and are applying Western approaches to recycling awareness and education. In Poland and Hungary, there are highly successful examples of recycling advisers going to door to door to raise awareness and offer advice on how to participate in recycling collections.

Raising public awareness with respect to broad waste management issues is a particularly important issue in the Russian Federation because, generally speaking, there is very low confidence in all aspects of waste management and in local government regulation and control. To help ensure continued public support and the successful implementation of solid waste management improvements, public participation is not just desirable; it is essential.

Waste: from individual ownership to public engagement

Whatever we are trying to achieve in the waste management field, from greater use and payment for waste collection services to improved participation in recycling, it must be communicated to the public, and they must be engaged as early as possible to build ownership, commitment and understanding.

Waste, as was explained earlier, is a social construction, so we must all engage with it. Waste is only waste because we no longer have a use for a specific material, product or item, and we must all take personal responsibility and ownership for the waste that we create.

We cannot afford to ignore the public any longer, and simply design systems for them to use. The public are not ignorant and know what they like – we must be brave enough to engage them in the decision-making process, and allow them to help evaluate options and ultimately determine the type of service they will get.

Clearly, the issues of public understanding, ownership and participation are vital to the successful evolution of the waste-management sector, as everything we do in waste management starts and ends with people.

The future of the waste industry is that of a social science. We have had policy-led epochs, phases of technological evolution, and stages where management and control have been paramount. Now we are on the cusp of a social evolution in waste management, with the realisation that waste management is about people, is driven by people and is for people.

It is time that we develop this concept and more fully understand society's role in improving waste management.

References

The Aarhus Convention is available at www.unece.org/env/pp/welcome.html.

'Best Practice Guidelines on Public Engagement for the Waste Sector' can be viewed at www.the-environment-council.org.uk.

Brook Lyndhurst Study *Household Waste: Attitudes, Behaviour and Opinions*, Vols I and II are available from the Resource Recovery Forum (www.residua.com).

Environment Agency. *Consensus Building for Sustainable Development* Environment Agency, UK, 1998.

EPA, Program for Community Problem Solving. *Involving Citizens in Community Decision-Making* (Washington, DC: EPA, 1997).

ERM. *Strategic Planning Guide for Integrated Solid Waste Management*, World Bank, 2001 (available on CD or from www.worldbank.org).

GLA Guidance Report on 'Best Practice in Doorstepping Campaigns' (can be accessed at www.london.gov.uk/mayor/strategies/waste/doorstepping.jsp).

Institute for Public Policy and Research Public Involvement Programme (can be accessed at www.pip.org).

Petts, J. 'Evaluating the Effectiveness of Deliberative Processes: Waste Management Case Studies', *Journal of Environmental Planning and Management*, 2001, 44(2), 207–26.

Renn, O., Webler, T. and Widemann, P. *Fairness and Competence in Citizen Participation* (Dordrecht: Kluwer Academic, 1995).

Swiss Centre for Development Cooperation in Technology and Management has useful information at www.skat.ch.

United Nations Development Programme has many case studies on community participation at www.undp.org.

WRAP National Recycling Communications Campaign (can be accessed at www.wrap.org.uk and www.recyclenow.com).

Section 2:
The Economics and
Finance of Waste

3

The economics of waste management: applying the equality principle

William Hogland
School of Pure and Applied Natural Sciences, University of Kalmar and
Department of Engineering, Physics and Mathematics, Mid Sweden University,
Sweden

Jan Stenis
Department of Engineering, Physics and Mathematics, Mid Sweden University,
Sweden

The paradigm shift

A new approach to the problem of waste is needed. Otherwise, the process of achieving environmentally sound industry may be unacceptably slow. The paradigm shift that is argued for here involves equating industrial waste with normal products in terms of the allocation of revenues and costs, an approach that is termed the 'equality principle' [1].

This approach is the basis of the following discussion. The waste fractions studied are regarded as a company output, as mathematically expressed in formula ① below. This formula is used to allocate the revenues and costs of a certain waste fraction by multiplying them by splitting them into their proper proportions. The current waste fraction thus represents a new, substantially larger, cost, a so-called *shadow cost* or *shadow price*, which, if fully taken into account, imposes very strong financial incentives to reduce drastically the waste in question. Thus, we have the formula,

$$\frac{A}{B + C} \qquad ①$$

where A = the quantity of a certain waste fraction produced, B = the quantity of normal product output, and C = the sum of the quantities of all the different waste fractions produced.

Of course, to apply formula ①, a suitable production or administrative unit must be defined, depending on the circumstances. Formula ① represents the financial implications of the equality principle and is termed 'the model for efficient use of resources for optimal production economy' (EUROPE) [2].

The EUROPE model gives PF = proportionality factor:

$$PF_x = W_x / (TP + TW) \qquad ②$$

(compare formula ① above) where

W_x = waste bad no. x; units: tonnes, litres, monetary value, etc.

TP = total product (P) output goods = $\Sigma P_x = P_1 + P_2 + ... + P_n$ ③

TW = total waste output bads = $\Sigma W_x = W_1 + W_2 + ... + W_n$ ④

This cost-allocation principle is useful internally as the best practice of redistributing costs associated with waste between different departments of a company. This results in a form of competition between different production units, enhancing environmental improvement and profit. This gives companies an incentive to reduce waste in order to improve their product estimates. This improvement incentive affects product cost estimates, for example; thus, it affects budgets and forecasts used as a basis for loan applications, for example, and therefore information to company shareholders.

Cost-benefit analysis

Method of overhead rates based on normal capacity

The first method to be discussed in terms of ways to estimate the 'true' internal cost of waste fractionation, in terms of the view presented, is that of *overhead rates based on normal capacity*. The estimated costs are allocated as follows:

TC/item = (estimated VC/calculated quantity of items) +

(estimated FC/normal quantity of items) ⑤

where TC = total cost, VC = variable cost, and FC = fixed cost.

This study proposes that, mathematically, allocation is achieved by multiplying PF_x by the first term in formula ⑤. Moreover, PF_x is used to allocate costs to a particular fraction through multiplication either by the total FC, so as to obtain the FC for the waste fraction in question, or by the second term appearing in formula ⑤, so as to obtain the FC per item or unit of the waste fraction considered.

Average cost estimation method

The second method that can be used in considering a company producing one product only, is simply to divide the total cost for the period in question by the total production during that period, resulting in the cost per tonne, litre, etc.

This study proposes that when applying the average cost estimation method, the cost of a given waste fraction x is determined by multiplying PF_x by the actual or budgeted average cost for the period in question. However, this will result simply in a cost of the waste per unit of the current total amount of waste in question. In a construction waste management context, the *average cost estimation method* has been found the best to apply.

Absorption costing method

The third method of estimation to be considered in connection with the separation of waste fractions is the *absorption costing method*, also used here in a somewhat modified fashion. This method involves a step-by-step analysis of the contribution of the separate costs to the final cost units, taking into account the following:

- the distribution of direct costs in the final cost units;

- the distribution of indirect (overhead) costs in the sub-organisations involved (such as departments);

- the distribution of the costs of the sub-organisations involved in the final cost units.

Estimates for a given product are made as shown in Table 1.

Table 1. Basic set-up for the absorption costing method

	DM
+	MO (= DM $\times$ absorbed indirect material costs rate)
+	DL
+	PO (= DL $\times$ absorbed production overhead costs rate)
=	production costs
+	AO + SO (= production costs $\times$ S, G and A rate)
=	total cost

where DM = direct material costs; MO = material overhead costs; DL = direct labour costs; PO = production overhead costs; AO = administrative overhead costs; SO = sales overhead costs; absorbed indirect material costs rate = $(MO_{Total}/DM_{Total}) \times 100$ (%); absorbed production overhead costs rate = $(PO_{Total}/DL_{Total}) \times 100$ (%); S, G and A rate = rate of absorbed general administration and marketing and sales overhead costs; i.e., the S, G and A expenses rate = (S, G and A$_{Total}$/total production cost) $\times$ 100 (%).

This study proposes that the possibility of using this method in connection with the assessment of the profitability of separating waste fractions lies in multiplying PF_x by 'total cost' in Table 1.

Activity-based costing (ABC) method

The *activity-based costing (ABC) method* is the fourth method to be considered in connection with waste fraction separation, again used in a modified form. If many of the costs arise from factors that are non-volume-based, the ABC method is clearly applicable. This is a method in which the costs of each activity in which an organisation is engaged are determined and then linked to the products, services or other objectives with which costs are associated. The aim is to trace costs to products or services instead of arbitrarily allocating them [3]. Since costs are often linked to the number of transactions involved in the activity in question, ABC is also called *transaction-based costing*. This study proposes that, in *this* case, using mass as the allocation basis, the proportionality factor K given in formula ⑥, which is similar to formula ①, should be multiplied by the total cost for a certain single product to obtain the cost to be allocated to the waste from the related component in question.

$$K = \frac{\text{Mass of the waste from production of a single component}}{\text{mass of the product } + \text{ total mass of waste from producing the product}} \quad ⑥$$

Contribution margin analysis

Contribution margin analysis (CMA) is a method of fundamental importance in the business context. It involves the assumption that, within certain limits, the fixed cost of a product is basically independent of the number of units manufactured or sold, and only the variable cost changes [4].

In contrast to cost-benefit analysis, CMA is used mainly for short-term decisions concerning the resources available at the time. It is usually a question, therefore, of making use of existing plants and the existing workforce in as profitable a way as possible. The crucial issue is whether the income from the waste fraction in question fully covers the fixed cost connected with it. If it is assumed that sufficient manpower is available, the decision of whether to commercialise a given waste fraction can be facilitated by assessing the contribution margin connected with it, in the manner shown in Table 2.

Table 2. Scheme for estimation of the contribution margin, shown here for the fraction of waste sold (cf. Horngren *et al.* [5])

	Income from sale of the fraction sold
–	Variable cost of the fraction sold
=	Contribution margin covering the fixed cost
–	Specific fixed cost of the fraction in question
=	Contribution margin after deduction of costs traceable to the fraction =
	= Operating income

This study proposes that if total revenue (TR) for a particular fraction cannot be determined, the recommended procedure is to implement formula ①, in a manner similar to that for FC and VC above. If a positive value is obtained in the bottom line, this generally means that the waste fraction in question should be turned into a product and not simply be dumped or discarded.

Assessment of the specific income and the specific fixed and variable costs, which is an important step, can be carried out according to traditional economic theory. The problem can be described as follows [4]:

$$TC = f(x) = FC + VC = FC + k_1 \times x, \text{ where } k_1 = (dy/dx) \text{ for VC} \qquad ⑦$$

$$TR = f(x) = k_2 \times x, \text{ where } k_2 = (dy/dx) \text{ for TR} \qquad ⑧$$

where

TC = total cost, FC = fixed cost, VC = variable cost and TR = total revenue.

This study proposes that this assignment involves multiplying PF_x by the total FC to obtain the FC for the waste fraction in question. If the VC directly attributable to a given fraction cannot be determined, costs can be allocated, just as for the FC above, in proportion to the weight or volume of the fraction, or to the amount of raw material or time consumed in producing it, by multiplying the respective VC by PF_x. If TR for a particular fraction cannot be determined, the recommended procedure is to implement formula ①, in a manner similar to that for FC and VC above.

After FC, VC and TR have been obtained for a given waste fraction, the operating income (or contribution margin), is estimated, for example, as income per unit of waste produced. This study proposes that this operating income be then incorporated into current wastes, after all relevant internal estimates have been made of short-term character, such as those for product costs. In this way, the existence of the waste fraction in question affects the estimate of the desired operating income.

Although applying the allocation principles contained in formula ① to the five methods described above redistributes the cost of regular products to waste – which does not necessarily result in an increase in the *total* cost for the company involved – it does not directly link the avoidance of waste with the incentive to reduce the total cost, as specified in the consolidated profit and loss account used for business purposes. *Weights* can be applied, however, to adjust the costs connected with a particular type of waste to its environmental impact, based on scientific evidence and/or in terms of overall societal aims. What can be termed 'environmental shadow prices' should therefore be used in combination with the cost allocation principle in defining environmental standards.

Polluter-pays principle

A commonly suggested way to cope with the pollution aspect of the waste problem, is to apply the *polluter-pays principle* (PPP); that is, to let the polluter bear the cost of the pollution-prevention and control measures that he has caused, the latter being measures decided by public authorities to ensure that the environment is protected [6].

At the Rio de Janeiro summit in 1992, the United Nations Conference on Environment and Development (UNCED) stated: 'Governments… should apply the PPP whenever appropriate… through setting waste management charges at rates that reflect the costs of providing the service and ensure that those who generate the wastes pay the full cost of disposal in an environmentally safe way' [7]. This study proposes that, in an internal business economic context, a first step to apply this principle would be to allocate all the necessary costs for making the production process environmentally friendly in a company – which is called *the environmental adjustment costs* – to the residual waste products involved. This additional allocation can be in relation to the weight, volume or value of the waste or time required to handle it.

This study proposes that $PF_{x'}$ when multiplied by *the environmental adjustment costs* that accrue, yields the costs connected with waste that are referable to a particular industrial activity with environmental repercussions. One needs to decide upon a suitable production or administrative unit to which PF_x is to be applied. This can be the entire company, separate divisions or workshops, individual machines or any other level of the production system. Environmentally related shadow price costs are allocated to the waste that is produced and are incorporated into the internal cost calculations of the company so to induce corporate waste-reducing incentives leading to cleaner production processes.

Waste management for sustainable production view revision

This study shows how the principle of equating industrial waste with regular products in a business sense can be applied to traditional cost-benefit methods as a financial basis for environmentally friendly waste management. Results of case studies carried out show that, in terms of the principle of the 'true' internal shadow price, the cost of (solid) waste is substantial, the exact costs obtained differing according to the method employed. The applicability of such methods is indeed indicated by the small variations in the estimated total cost of the waste fraction in question in calculations made without employing environmental impact weighting.

There appear to be no specific obstacles to applying the proposed equality principle to either a bulk industry or the manufacture of technically complicated products. The methodology suggested can be assumed, therefore, to be generally applicable to all manufacturing companies producing waste. Hopefully, the present study also will encourage the development of alternatives to traditional taxation practices in the environmental context.

Through creating a direct link between budgetary costs and the production of waste, that is, the redistribution of costs between normal products and waste, economic pressure is exerted on the industry to introduce environmentally friendly measures that are as effective as possible in reducing waste at source; these measures also tend to enhance production efficiency and thus profitability. So that forecasts, company budgets and consolidated profit-and-loss accounts for external use that companies present will be affected in a way that 'punishes' the excessive occurrence of waste and the failure to utilise it productively, both appropriate official recommendations and voluntary environmental agreements regarding the assessment of industrial wastes are needed. Such developments may be necessary to force industry to act in a manner truly in accordance with the ideal of improving the sustainability and productivity of resource use.

The study will contribute to effecting a change in the perceived status of industrial waste through emphasising the financial implications of the costs and revenues involved and the use of shadow prices. It is hoped that the basic approach advocated here will be adopted by industry generally, and will be accompanied, to a certain extent, by a paradigmatic shift towards sustainable production in long-term development, regarding the perception of industrial waste, together with better ways of reducing waste and of utilising waste fractions unavoidably produced. In this way, a concrete contribution can be made to the fulfilment of the waste-reducing ambitions of, for example, the central European authorities, the overall goal being the achievement of more sustainable development.

Conclusions and recommendations

The effort described here represents a shift of views within the field of waste management. This shift is consistent with the concept of sustainable development. Moreover, the findings represent general environmental and financial advantages for society at large.

It is concluded that the lessons learned point to the partly fruitful ability to modify commonly used cost-benefit estimation methods and contribution margin analysis, including the polluter-pays principle (PPP), in a practical industrial and, to a certain extent, a practical construction waste management context by the use of the equality principle.

Summarised, the major results and outcomes of the work are as follows:

1) presentation of an alternative way of looking at industrial waste in a business context

2) introduction of the EUROPE model for assigning costs to industrial waste to be used in conjunction with the introduced equality principle to provide long-term recommendations regarding waste involving the use of 'environmental shadow prices'

3) presentation of how the polluter-pays principle can be incorporated into the financial accounts of a manufacturing company by the employment of

the equality principle in conjunction with the introduced concept of the 'environmental adjustment cost'

4) elaboration of a principle for estimation of waste-related company costs and revenues

5) implication of industrial and constructional cost-saving incentives

6) reduction of waste at the source, leading to less waste produced

7) extended environmental good will from adequate waste management

8) enhanced status of waste due to a new way of regarding such waste as being equivalent to regular products in financial terms

9) proposal for a shift of waste-management paradigms.

Summarised, the recommendations for the future are as follows:

1) As regards *general industrial conditions*, commonly used cost-benefit methods, the contribution margin method and the polluter-pays principle approach are recommended to be applied in a modified way through the use of the EUROPE model, the basis being the equality principle introduced.

2) As regards *construction management conditions*, the average cost-estimation method is recommended to be applied in a modified way through the use of the EUROPE model, the basis being the equality principle introduced.

References

[1]   Stenis, J. *Industrial Waste Management Models – A Theoretical Approach*. Department of Construction and Architecture, Division of Construction Management, Lund Institute of Technology, Lund University, (licentiate dissertation), 2002.

[2]   Stenis, J. *Industrial Management Models with Emphasis on Construction Waste*. Department of Construction and Architecture, Division of Construction Management, Lund Institute of Technology, Lund University, Department of Technology, University of Kalmar, (doctoral thesis), 2005.

[3]   Johnson, H.T. and Kaplan, R.S. *Relevance Lost. The Rise and Fall of Management Accounting* (Cambridge, MA: Harvard Business School Press, 1987).

[4]   Frenckner, P. and Samuelson, L.A. *Produktkalkyler i industrin* (*Product Estimation in Industry*) (Stockholm: Sveriges Mekanförbund, Mekanförbundets Förlag, 1989) (in Swedish).

[5]   Horngren, C.T., Sundem, G.L. and Stratton, W.O. *Introduction to Management Accounting* (Prentice-Hall International, 2002).

[6]   OECD Environment Directorate. *The Polluter-Pays Principle. OECD Analyses and Recommendations*. OCDE/GD(92)81, 1992.

[7]   UNCED. *Report of the United Nations Conference on Environment and Development*, Rio de Janeiro, 3–14 June 1992, ch. 30.

4

The economics of carbon finance – financing landfill gas projects in the waste sector

Jelmer Hoogzaad, Charlotte Streck, Adriaan Korthuis, Robert O'Sullivan and Sandra Greiner
Climate Focus, Rotterdam, The Netherlands

This chapter discusses a key dimension of 'WASTEnomics' – that of turning the liabilities of a waste landfill project into an asset through carbon finance.

Introduction

Since the Kyoto Protocol came into force in February 2005, the total value of the global regulated carbon markets grew from 9.6 billion euros in 2005 to 22.5 billion euros in 2006. The Kyoto Protocol defines two project-based trading mechanisms: joint implementation (JI), which allows the crediting of carbon credits generated by projects in developed countries and countries with an economy in transition, and the clean development mechanism (CDM), which involves developing countries in the compliance system of the Kyoto Protocol [1]. At current market prices, the value of carbon credits from JI and CDM projects that are in advanced stage of development amounts to approximately 21 billion euros. Carbon credits from projects that aim at the extraction and combustion of landfill gas ('LFG') make up about 9.3 per cent of the JI and CDM market [2]. Over 240 advanced CDM LFG projects in 44 countries [3] indicate that JI and the CDM have developed into a relevant source of financing for projects involving methane extraction and destruction of LFG.

This chapter will give some background on the Kyoto Protocol's carbon-trading mechanisms and share practical market experiences that are relevant to LFG project developers, carbon buyers and financial institutions. The first four sections describe the Kyoto Protocol, its trading mechanisms, and the CDM and JI project cycles. Further sections include an analysis of LFG projects under JI and the CDM, the effect of carbon finance on project feasibility, and the contractual context of LFG carbon agreements.

The Kyoto Protocol and project-based emission trading [4]

Russia's ratification of the Kyoto Protocol in November 2004 levelled the path for the entry into force of the Kyoto Protocol. Among the most innovative features of the Kyoto Protocol is the 'flexible mechanism' [5]. The treaty's compliance system, which is based on capping emissions of industrialised countries while giving them the opportunity to explore least-cost emission-reduction options through the exchange of tradable emission quota, is unprecedented in global environmental policy.

The Kyoto Protocol establishes emission limitation and reduction targets for countries that ratified the Kyoto Protocol and are listed in Annex I of the UN Framework Convention for Climate Change (UNFCCC) [6]. Annex I of the UNFCCC lists OECD countries (except Mexico and South Korea) and countries of eastern Europe. With global emissions at about 33,750 million tonnes of CO_2-equivalent emissions [7], Annex I countries were responsible for 54 per cent of the global emission of greenhouse gases (GHGs) in 1990. Although the USA is listed in Annex I, it is excluded from JI and the CDM since it has not ratified the Kyoto Protocol [8].

Developing countries are commonly referred to as non-Annex I countries. These countries have not committed themselves to cap their GHG emissions. By ratifying the Kyoto Protocol, they have, however, become eligible to host projects under the CDM.

The Kyoto Protocol foresees the trading of assigned allocations of carbon rights ('assigned amount units') among Annex I parties. With the CDM and JI, it further defines project-based trading mechanisms. Credits created by the CDM and JI result from projects that reduce emissions below a 'baseline' (Figure 1). The baseline describes how the emissions would develop in the absence of the CDM or JI project. Project-based emissions trading enable countries and entities that operate under an emission cap to achieve emission reductions in a more cost-efficient manner. Allowing countries and governments to generate emission reductions where it is cheapest to do so can reduce the costs of mitigating global climate change.

Governments from Annex I countries can buy certified carbon credits from CDM or JI projects. They can also authorise companies to acquire or sell these credits. The cash flow generated by these purchases gives projects that reduce GHG emissions a competitive advantage. In some cases, projects generate no other income besides the revenues from the sale of credits [9]. The CDM has demonstrated that it is an operational trading mechanism ever since the first credits were issued by the UNFCCC in October 2005 [10]. Apart from reducing emissions, the mechanism aims to stimulate investments in developing countries and create technology transfer [11].

Within two years after the Kyoto Protocol entered into force, the trading mechanisms developed into a driving force behind investment in the reduction of GHG emissions. The carbon market gave a particular boost to those projects that involve GHG with high global warming potentials, such as CH_4, N_2O, and HFC_{23}.

The JI and CDM project cycles

To ensure the environmental integrity of the trading mechanism and the credibility of the credits, the Parties to the Kyoto Protocol agreed on a complex set of requirements and conditions that any CDM or (to a lesser extent) JI project has to fulfil. The first step in the development of a CDM or JI project is the elaboration of a Project Design Document (PDD). The PDD contains the calculation of the emission reduction expected to be created by the project activity in comparison with a baseline scenario. The baseline scenario describes the development of the GHG emissions in the absence of the project. The amount of credits a project developer can claim equals the difference between the baseline emissions and the actual emissions of the project (see Box 1). In addition to estimating the emission reduction, the PDD also contains an additionality test which demonstrates that the project would not have occurred without the CDM. This test is also an element of a JI project but may be applied in a less stringent manner [12].

After estimating the emission reductions in the PPD an independent third party, a validator, assesses whether the PDD meets the requirements of the CDM or JI. The validator also checks whether the host country government has approved the project.

In the case of the CDM, if the validator determines that the PDD meets all requirements, the project can be submitted for registration as a CDM project. Registration happens through a decision of the Executive Board (EB), the Kyoto Protocol body that supervises the implementation and operation of the CDM. The EB checks whether the validator assessed the project according to the CDM modalities and procedures. Once approved by the EB, projects are 'registered' as a CDM project and can start generating CERs.

Box 1. Developing a PDD

The Project Design Document provides an estimate of the emission reductions that a project will generate. The estimate is based on the difference in emissions between the project scenario and the baseline scenario. The actual reduction measured during operation of the project may differ from the forecast in the PDD. It is relatively easy to monitor the emission reductions from a LFG project activity because the reduction equals the amount of methane combusted. On the other hand, it is relatively complicated to forecast the amount of methane that will be extracted and combusted.

Making an *ex ante* estimate for the PDD requires the use of decomposition models. The amount of methane emitted by a landfill depends on factors such as the amount of organic material in the landfill, the permeability of the landfill to methane and outside air, temperature and moisture content.

A number of panels support the EB. An important panel is the Meth Panel, which prepares recommendations on proposals for new baseline and monitoring methodologies. The EB needs to approve methodologies that project developers use for the calculation and monitoring of emission reductions. Methodologies for the calculation of emission reductions from LFG projects were among the first approved by the EB [13].

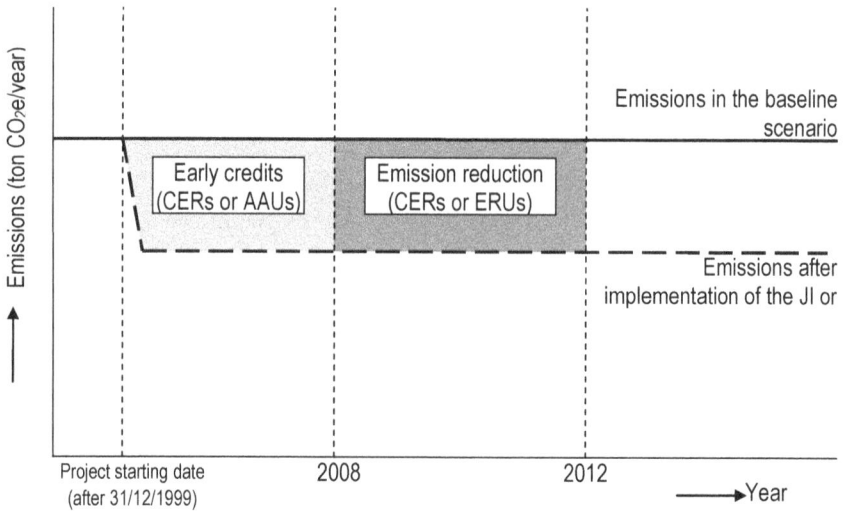

Figure 1. Emission reductions due to the implementation of a CDM or JI project calculated against a baseline scenario.

The project developer must monitor the actual emissions from the project and submit monitoring reports to an independent third party for verification. After confirming that the emission reductions claimed by the project developer actually occurred, the verifier can ask for the issuance of carbon credits. For the CDM, these credits are called certified emission reductions (CERs) and for JI, emission-reduction units (ERUs).

An important difference between JI and the CDM is that ERUs are issued by the hosting Annex I country and deducted from the country's allocated assigned amount. CERs are from non-Annex I countries without assigned amounts and are issued by the EB.

For JI, there are two different project cycles, often referred to as track 1 and track 2. Track 2 applies if the host country does not meet all accounting requirements that make it fully eligible for hosting JI projects. The track 2 procedure is similar to the CDM procedure. JI track 2 is supervised by the international JI Supervisory Committee, the JI version of the CDM Executive Board. Under track 1, the host country is authorised to validate projects and verify emission reductions without the involvement of the Supervisory Committee. A host country can then decide on the criteria it uses for the assessment of JI projects. Track 1 offers host countries the opportunity to make JI significantly simpler and reduce transaction cost. However, project developers can only use track 1 in countries that have a well-functioning system for the accounting of emissions. Whether countries meet the requirements for a proper accounting of emissions depends on the evaluation of countries' progress in the implementation of the relevant systems. For most Annex I countries, such evaluation will have taken place by the end of 2007.

Supply and demand on the carbon markets

The future demand and supply on the carbon markets is determined by a number of unpredictable factors, including the weather, the allocation of allowances under the EU Emission Trading System, the prices and availability of oil and natural gas, and the demand for electricity and heat. The total compliance gap of Japan is estimated at 1 billion, that of Canada at 1.2 billion and that of the EU-15 at about 1.5 billion. The different countries in the EU-15 together have already planned to use the flexible mechanisms to purchase over 0.5 billion carbon credits, a purchase to which they allocated 2.8 billion euros [14].

On the supply side, the total amount of credits from JI and CDM projects under or past validation exceeded 2 billion tonnes of projected CO_2e emission reductions in January 2007. This supply will fill a part of the Annex I compliance gap. In addition, EU members may engage in trade of AAUs with each other.

The Kyoto Protocol compliance period ends in 2012. The World Bank estimated the aggregated value of carbon markets for 2006 to be US$25–30 billion, of which 11 per cent came from JI and CDM transactions [15]. Whether this value is sustained for the years until 2012 depends to a large extent on the outcome of post-Kyoto discussions. If countries agree on commitments after 2012, the incentive to comply with commitments under the Kyoto Protocol will be stronger. The demand for CDM or JI projects will depend on whether countries engage in direct emission trading (not project-based), whether Canada makes an effort to meet its Kyoto target, and whether countries agree on post-Kyoto commitments.

LFG projects under CDM and JI

In this section, we discuss the potential of carbon finance in waste landfill and LFG projects. The anaerobic decomposition of organic material in landfills can generate large amounts of methane. Usually this methane is emitted to the atmosphere. Since methane is a GHG as well as a fuel, some landfill managers decide to extract it and destroy or even use it. To extract the methane, they install suction wells in the body of the landfill. Vacuum pumps ensure that low pressure in the wells creates a flow of landfill gas from the landfill body. A pipeline system then transports the landfill gas to a central point where the methane in the landfill gas can be combusted. The volume percentage of methane in the landfill gas may vary. Depending on the caloric value of the landfill gas, it can be flared or used for heat or power production.

LFG projects have a prominent position among CDM and JI projects. In December 2007, of 3042 CDM and JI projects under validation, 240 were LFG projects, of which 64 were registered. The potential GHG emission reduction of the LFG projects in the pipeline amounts to 239 million tonnes [2] of CO_2-equivalent emission reduction, representing a market value of about 1.7–3.3 billion euros (at a price of 7–14 euros per tonne of CO_2e). The lines in Figures 2 and 3 indicate that although the share of credits from LFG project is decreasing, the total amount of credits from LFG projects is still increasing at a stable rate. The bars in Figure 3 confirm this by showing a relatively stable amount of credits from LFG projects developed each month.

At the time of writing this chapter (December 2007), about 7 per cent of the CDM projects under or past validation are landfill gas projects. These projects are responsible for 9 per cent of the total planned emission reductions under the CDM. The share of LFG projects in JI is 12 per cent, covering 6 per cent of the projected emission reductions. Figure 3 shows the share of CERs from LFG projects in the CDM market, including N_2O and HFC_{23} projects. These projects reduce the emissions of very potent GHGs, and only 15 of these projects are expected to generate about half the carbon credits from the JI and CDM markets until 2012 [2]. If we ignore the emission reductions of these exceptionally large projects, the share of LFG projects doubles [2].

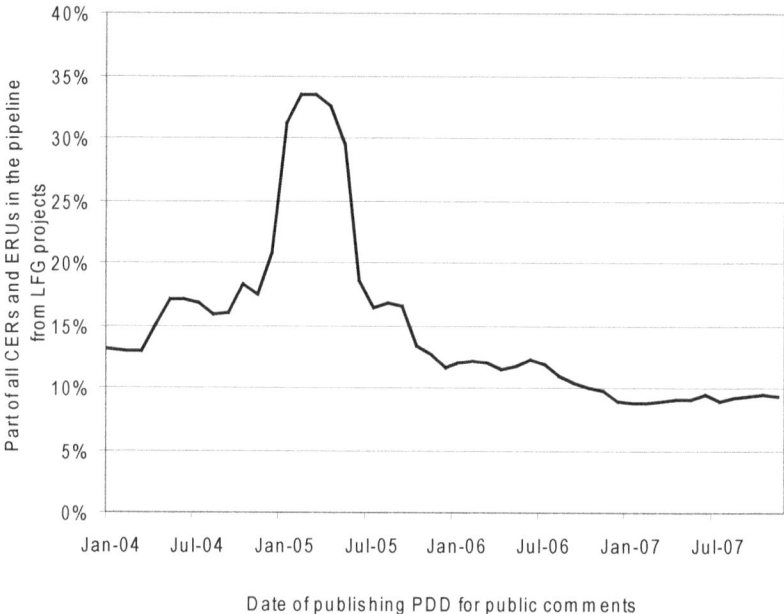

Figure 2. Share of CERs and ERUs from LFG projects in total amount of credits from JI and CDM projects of which the PDD is published for validation.

While 75 per cent of all CERs is expected to come from Asia and the Pacific (75 per cent), 72 per cent of the LFG credits will be from landfills in Latin America, a third of which are in Brazil [16]. New PDDs for LFG projects are published for validation every month, in particular from projects in CDM countries. Over the last two years the accumulated number of credits in these PDDs seems to be stable. Although this is not solid evidence, it may indicate that the LFG project potential is not yet depleted because project developer continue to find new projects every month (Figure 3).

The possibility of trading of carbon credits from JI countries is limited. The Kyoto rules exclude the accounting for emission reductions by LFG projects that correspond to a legal obligation. EU Directive 31/1999/CE requires that EU member states reduce the amount of biodegradable material that is disposed in landfills. That limits the potential for the generation of carbon credits from LFG projects within the EU.

Estimating emission reductions before project implementation is necessary to develop a PDD and register the project as a JI or CDM project. LFG projects typically involve the destruction of methane that is actively extracted from a landfill with vacuum pumps. The amount of methane that would have been emitted without the project is considered equal to the amount that is captured and destroyed. There are five approved landfill gas CDM methodologies [17]. Three of them prescribe the use of the LANGEM first-order decay model developed by the US Environmental Protection Agency, whereas the other two allow for more flexibility on how to make an *ex ante* estimate of the methane production. The first-order decay model enables quantification of the amount of landfill gas produced over a certain period based on the amount of previous and future waste disposed, its organic material content, and a decay rate. The decay rate depends on the humidity and temperature of the landfill.

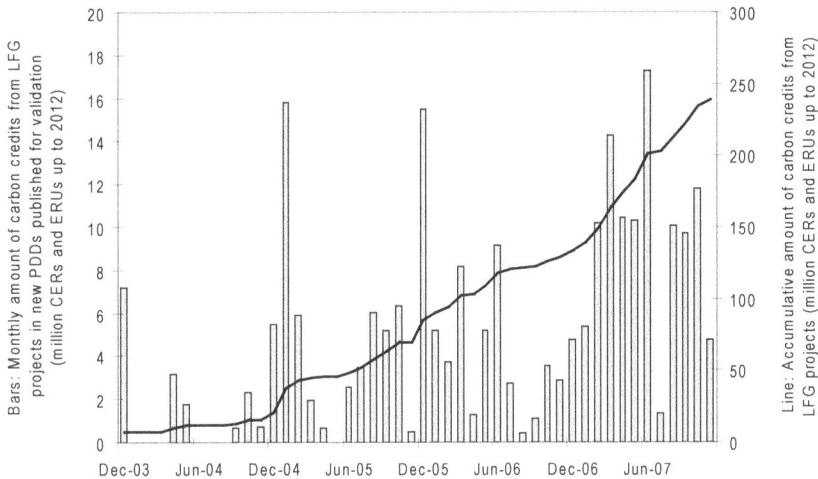

Figure 3. Total amount of credits expected up to 2012 from LFG projects of which the PDD is published for validation in a given month; the expected accumulative amount of carbon credits up to 2012 from LFG projects [2].

Future methane production and capture are difficult to estimate, and even in a single landfill the spatial variation in production can be very large. LANDGEM [18] is based on typical waste compositions in the USA corrected for differences in the amount of inert material. LANDGEM was developed to bring most of the large US landfills into an air-quality regulatory programme. Its application as health and safety standard leads to a conservative overestimation of gas generation. Such overestimation leads to over-dimensioned gas extraction and utilisation equipment but also to overly optimistic emission reduction delivery obligations [19]. Landfill experts have therefore questioned whether LANDGEM is an appopriate model for gas generation. In addition to the LANDGEM model, project developers also use a model adopted by the Intergovernmental Panel on Climate Change (IPCC) [20]. Another example of a gas-estimation model is the one developed by the Dutch research institute TNO, which was later improved by the University of Wageningen. The new version is also a first-order decay model; however, it takes the waste composition into account. It calculates the methane emissions by the biodegradable organic carbon content of various waste fractions.

Apart from the difficulties in estimating future methane production, the amount of methane actually captured can also be difficult to estimate. Studies indicate that, given the oxidation rates of atmospheric methane, the following methane recovery rates are reasonably achievable with active methane recovery:

- 35 per cent for an operating landfill;
- 65 per cent for a temporary covered landfill;
- 85 per cent for a landfill with clay final cover;
- 90 per cent for a landfill with geomembrane final cover [21].

The performance of implemented CDM LFG projects indicates that PDDs generally tend to overestimate the amount of emission reduction from the projects. In 2005, the first credits from LFG projects were issued by the UNFCCC based on verified monitoring reports. These reports provide information on the performance of LFG projects compared to the emission-reduction estimates in the PDDs. Monitoring reports from the 13 most advanced LFG projects show that they only deliver 35 per cent of the amount of credits estimated in the PDDs. Possible reasons for the gap between PDD estimates and the eventually monitored emission reductions are as follows:

- bad modelling;
- inefficient design of the extraction system;
- delays in project implementation;
- landfill practice that does not favour rapid extraction;
- deliberate overestimation.

Landfills that are closed to further waste usually generate a decreasing amount of methane. Project developers can choose to decommission part of the flaring or power-generation capacity within 5–10 years after closure of the landfill. Landfills with an increasing annual amount of waste tend to generate an increasing amount of methane. This allows a gradual upscaling of the methane combustion capacity and also an increasing annual amount of credits to be generated [22]. Some landfill gas projects already start constructing horizontal gas extraction wells when the landfill is still in operation. This way, the total amount of gas extracted from a landfill can increase significantly.

Impact of carbon finance on LFG projects

Compared to renewable energy and energy efficiency projects under the CDM, the investment costs of LFG projects are relatively low. For projects at closed landfills, the following investments need to be made:

- drilling gas extraction wells in the landfill;
- connecting the wells to a suction system with vacuum pumps;
- destroying the gas in a combustion system, usually by flares or a gas engine with generator.

The extraction wells in closed landfills are often vertical. In operating landfills, vertical wells would make it difficult to add waste; therefore, operating landfills often have horizontal extraction wells to allow simultaneous addition of waste and extraction of gas from the waste that is already accumulated. If the layer of new waste is thick enough, new horizontal extraction wells can be constructed.

Box 2. Milestones in the development of CDM or JI projects

Typical milestones in the development of a CDM or JI project are as follows:

Financial closure. Some forward carbon contracts contain a clause that they only enter into force once the project reaches financial closure. Depending on the reputation of the buyer of the carbon credits and the terms and conditions of the contract a signed forward contract may help project developers attract financing for the project.

Host country approval. In general obtaining approval of the Annex I country where the buyer is located is relatively easy and efficient. Getting approval of the host countries can be more difficult. Some countries are yet to develop and implement an approval procedure for JI or CDM projects. Russia is an example. Other countries formulated specific requirements for CDM projects. Most host countries give priority to projects in which LFG is used rather than just flared.

Registration. The international registration of CDM projects is well tested and LFG projects have rarely confronted insurmountable problems. The project cycle and approval procedure for JI has been implemented in 2006. In March 2007 the first JI project was registered on the UNFCCC web-site.

Selling the CERs or ERUs. There are two options for selling CERs and ERUs – they can either be sold under forward contracts before they are generated or they can be sold on the spot market after they have been generated. Forward contracts guarantee a predictable cash flow. Selling CERs on the spot market on the other hand holds the promise of higher carbon prices.

Sometimes, once the landfill is closed, vertical wells are also implemented to optimise extraction rates.

The investment costs depend on whether the project will generate power or merely flare the LFG. Table 1 shows the results of a financial analysis by the World Bank for LFG projects that differ in size. In its analysis, the World Bank assumed the following:

- The carbon price is US$6 per CER (this can be considered conservative since current prices are 2–3 higher) [23].

- The projects are developed with 25 per cent equity from the project developer.

- Only the CERs generated up to the end of 2012 are sold.

- Power can be sold against local grid prices [24].

Selling carbon credits generated by an LFG project make an important and often decisive contribution to the feasibility of LFG projects. Some LFG projects use the extracted gas to generate electricity or heat. Generating power from LFG projects raises the investment costs significantly. Consequently, unless the power feed-in tariff is very high, projects are economically more attractive if they only flare the landfill gas. Table 1 shows examples of LFG extraction projects.

Table 1. Financial indicators of different projects with or without power generation [22, 24]

Landfill and location	Landfill size (Mtonnes)		Project type (with generation capacities until 2012)	Investment (mln US$)	Annual O$M costs (mln US$)	NPV (mln USD)	IRR (%)
	2005	2012					
Queretaro landfill, Queretaro, Mexico	0.3	0.4	Flaring only	1.10	0.15	0.69	48
			2.1 MW power generation	3.86	0.43	1.13	22
The El Combeima landfill Ibague, Colombia	1.1	1.1	Flaring only	1.22	0.08	-1.20	n.f.
			Power generation	n.f.	n.f.	n.f.	n.f.
Santa Tecla landfill, Gravataí, Brazil	1.8	2.0	Flaring only	0.92	0.09	0.39	77
			1.0 MW Power generation	2.15	0.23	-0.29	n.f.
The Chihuahua landfill, Chihuahua, Mexico	3.5	6.5	Flaring only	1.00	0.11	1.16	79
			2.1.MW Power generation	3.18	0.39	2.30	43
El Carrasco landfill, Bucaramanga, Colombia	5.0	6.9	Flaring only	1.12	0.18	0.73	85
			1.0 MW Power generation	2.43	0.32	0.27	21

Landfill and location	Landfill size (Mtonnes) 2005	2012	Project type (with generation capacities until 2012)	Investment (mln US$)	Annual O$M costs (mln US$)	NPV (mln USD)	IRR (%)
			Flaring only	1.71	0.12	3.00	111
Huaycoloro landfill, Lima, Peru	6.3	11.8	6.0 MW Power generation	7.01	0.92	2.58	44
Montevideo landfill,			Flaring only	1.60	0.12	3.10	148
Montevideo Uruguay	7.6	over 10.5	2.0 MW Power generation	5.33	0.40	0.36	13
Muribeca landfill, Pernambuco,			Flaring only	3.54	0.35	6.39	150
Brazil	11.2	14.4	7.4 MW Power generation	10.85	1.33	3.49	34
The Gramacho landfill, Rio de			Flaring only	5.89	0.30	12.9	284
Janeiro, Brazil	29.1	29.1	10.0 MW Power generation	15.52	1.62	9.20	77

The investment costs include preparation of the PDD, validation and registration. n.f.: not feasible.

In some cases, project developers can also sell credits for reductions generated before 2008 and after 2012. The prompt-start provision of the CDM allows projects implemented after 2000 to claim credits generated before 2008 and before the project is registered. Some JI host country governments also allow JI projects to claim emission reductions generated before 2008. Based on verified emission reductions, the host country government may transfer assigned amount units (AAU) for these early emission reductions. The project activity is considered to have 'greened' the AAU credits under this transaction. Extending the number of years a project can sell credits will have a clear benefit on the financial feasibility of the project.

Sale of carbon credits

The undisputed title to the future emission reductions is a condition for the sale of carbon credits. In marketing CERs or ERUs from LFG projects, it is important to ensure that the seller has the full, beneficial title to the CERs. Most LFG projects have a number of different project participants, a factor that complicates the ownership structure and makes the success of the project dependent on their ability to cooperate. Apart from the investor in the LFG extraction and incineration system, there is usually also a (commercial) landfill operator and a municipality that owns the landfill. The buyer will perceive an unclear ownership structure as a risk to delivery and is likely to refrain from entering into an agreement.

Provided the title of the carbon credits can be assigned, the seller can decide to sell the expected carbon credits either as forward contract before they are generated or on the spot market after they have been generated.

Forward contracts are often referred to as emission reduction purchase agreements (ERPAs) [25]. The main difference between these two types of contracts is that for forward contracts the project developer agrees to sell the credits at an agreed price before they are generated (though payment is normally upon delivery of the credits). With spot contracts, the credits are sold only after they have been generated and are sitting in a registry account ready to be delivered. Because forward contracts sell credits for emission reductions that still have to be achieved, there is some uncertainty over whether or not the credits will be generated. This typically translates into a price discount for the credits. Project developers that enter into ERPAs early in project development typically cannot fully mitigate higher delivery risks and as a result receive lower prices for their credits. Contracting early can still be useful if the buyer is willing to make advance payments to develop the PDD and cover other Kyoto Protocol-related expenses.

Both forward contracts and spot contracts can generate additional hard currency income for project developers. Forward contracts can be used to help secure advance payments to contribute to the investment capital of the project, though many buyers also request security in the form of letters of credit or a bank guarantee for such advance payments.

For forward contracts, it is important either to estimate the amount of emission reductions delivered in the future in a conservative manner or refrain from offering guaranteed delivery amounts. Depending on the contractual arrangement between the seller and the buyer, under-delivery of carbon credits may allow the buyer to claim damages or demand delivery of replacement credits. To avoid this, sellers could opt for selling only a part of the credits in a forward contract and sell any excess of credits on the spot market once it is generated. This is particularly apt for LFG projects where the generation of credits is dependent on uncertain biological processes and where municipal owners of LFG projects may be unable to assume the same liability for under-delivery that other private companies may be able to assume.

There are a number of pricing options available under an ERPA, including fixed prices and indexed prices. Fixed-price contracts have a fixed price per credit over the term of the contract. This can be useful for the buyer and seller, as they each know in advance how much they can expect to pay and receive for the credits. However, if the market price of the credits increases or decreases over the term of the contract, one party will inevitably pay or receive more or less than the market price. This can be overcome by using indexed prices where the price paid per credit is calculated against the current market price, with some discount to reflect risks. This will ensure the buyer does not pay too much, and sellers do not receive too little for their credits, but the lack of a set maximum and minimum price also exposes the parties to market fluctuations and creates some uncertainty over long-term price liability and revenues.

Box 3. LFG extraction at the Gramacho landfill, Rio de Janeiro, Brazil

This landfill has been in operation since 1990 and covers about 140 ha. A pump test was undertaken to estimate the gas potential. With a planned content of 29 million tonnes of waste by 2005, the landfill could generate enough gas to operate a 10-MW reciprocating engine generator set, consuming about 6,060 m³/hour until 2012. In order to be able to combust all methane, additional gas-flaring equipment with a capacity of 22,950 m³/hour will be installed. In the period 2006–12, the landfill could generate 5.53 million tonnes CO_2e, making it a relatively large project.

Let us apply the low CER price of US$6, assume an equity investment of 25 per cent, and assume only sale of credits between 2006 and 2012. With power generation, the project will have a net present value (NPV) of US$9.2 million and an internal rate of return (IRR) of 77 per cent. Without power generation, the project financials will be enhanced to an NPV of US$12.9 million and an IRR of 284 per cent, mainly because the initial investment will be reduced from US$15.52 million (operation and maintenance US$1.6 million) to US$5.89 million (operation and maintenance US$295,000) [24]. The power baseline in Brazil is relatively low because most electricity in the country is generated by hydropower. The additional emission reduction achieved from the generation of electricity is thus low.

The UNEP Risø statistics indicate that out of the 80 LFG CDM projects under or past validation, 43 projects include power generation. In only 21 cases, the project developers decided to include power generation in the emission-reduction calculations.

Buyers offer carbon prices on the basis of the risk profile of the project. Four key price determinants are:

- delivery risk;
- counter-party risk;
- host-country risk;
- reputational risk.

The perception of the delivery risk correlates with the development stage of the project [26]. Examples of reputational risk include cases in which the landfill supports poor local communities. Covering and closing the landfill can affect local stakeholders (scavengers, for example) and deprive them of their only source of income. This needs to be addressed in the project design to ensure that the project has a positive rather than a negative effect on the local stakeholders. Dissatisfaction of neighbouring communities with the operations of the landfill can also negatively affect a carbon project.

The carbon market has started to move from a buyers' market towards a more balanced and competitive market. From 2001 until the end of 2004, there were only a few buyers that could pick and choose between projects. From 2004, many private buyers started actively looking for projects, giving sellers many more options. More buyers now try to secure projects at an early stage of development and establish forward contracts, and more landfill operators receive all-inclusive offers. These offers include project development, implementation, and the generation, verification and sale of carbon credits. This can be good for sellers that do not want to pay for these themselves, but it can also be a disadvantage, as support at an early stage typically corresponds to a lower price per credit. Some sellers that choose to develop the project themselves may be able to obtain a higher carbon credit price by waiting – as long as the carbon market price does not decrease during this time. Although this approach is sensible, it is only possible for those projects that do not require prepayment on the sale of credits or a carbon agreement to be able to reach financial closure. In addition to these two extremes, it is also possible to make combinations, selling a portion of the credits under a forward contract and keeping the rest for the spot market.

Box 4. Feasibility of LFG extraction from the Montevideo landfill, Uruguay

With 7.6 million tonnes of waste in 2005, the Montevideo landfill project is smaller than the Gramacho project discussed in Box 3. The landfill is expected to generate 4,000 m³/hour between 2006 and 2012, after which the amount will decline to 900 m³/hour in 2019. If the landfill gas prognosis is true, the landfill could fuel a 2-MW power plant.

At a CER price of US$5 and a power sales price of 0.027 US$/kWh, power generation is only feasible if the project can also sell CERs beyond 2012. If all gas is flared, the project has an NPV of US$3.1 million and an IRR of 148 per cent when selling CERs only until 2012.

Discussions on post-Kyoto obligations are continuing. At the moment, it is not certain that the CDM or JI will exist after 2012, and if they do, what commitments may exist for buyers. As a result, there is significant uncertainty over post-2012 demand and post-2012 prices, and most ERPAs only cover the CERs and ERUs generated until 2012.

Conclusion

Carbon finance brings financing and emission-trading opportunities to the waste sector. It turns the liability of the landfill into an asset. LFG projects have already reaped some of the financial benefits of the carbon markets. However, there are also pitfalls that project developers should bear in mind:

- There is a significant uncertainty in the amount of credits a project can generate. Project developers should keep this in mind when fixing delivery amounts in carbon agreements.

- Strong municipal involvement in most landfill gas projects can make agreements sensitive to political changes.

Project developers that do not need pre-financing could choose to offer their projects on the spot market and avoid discounts for the risk of forward contracts. However, there is a risk in waiting. Carbon prices are volatile and potentially total supply on the carbon market is larger than total demand. Current post-Kyoto negotiations add additional uncertainty to the future value of carbon credits. The post-Kyoto GHG emissions regime will have an effect on carbon prices.

For small landfills, the transaction costs of emission trading can be too high. In general, projects at larger landfills are more feasible. Power generation can be attractive for larger landfill, but, despite being an attractive energy source, using LFG for power generation often makes the project less financially profitable.

References

[1]   The JI countries with the largest project pipeline are Russia, Ukraine, Bulgaria, Poland and Romania. The main CDM countries are China, India, Brazil and South Korea.

[2]   Calculations based on Fenhann, J., *JI and CDM pipeline* (Roskilde: UNEP Risø Centre, December 2007). For estimation of the size of the market, we assumed a price of 10 euros per carbon credit. The total amount of credits in published PDDs of JI and CDM projects is 2.1 billion carbon credits.

[3]   The April 2007 *JI and CDM pipeline* from the UNEP Risø Centre only lists projects that are past validation or are still undergoing validation.

[4]   For an analysis of the carbon markets, see Kenny Tang, *The Finance of Climate Change* (London: Risk Books, 2005).

[5]   Kyoto Protocol, articles 3, 6, 12 and 17 (1997).

[6]   Under the Framework Convention for Climate Change (1992) countries committed themselves to reducing GHG emissions. The Kyoto Protocol (1997) went a step further by defining caps on national emissions between 2008 and 2012 and introduced three trading mechanisms. The Marrakech Accords, which were adopted by the first CoP/MoP of the Kyoto Protocol in Montreal in 2005, includes the implementation guidelines of the Kyoto Protocol flexible mechanisms.

[7]   The Netherlands Environmental Assessment Agency, press release: *Global Greenhouse Gas Emissions Increased 75% Since 1970* (Bilthoven, 2006).

[8]   UNFCCC, Kyoto Protocol: Status of Ratification, www.unfccc.org.

[9]   Some examples of projects whose main source of income is revenues from the sale of carbon credits are LFG and Coal Mine Methane projects in which the methane is flared and projects in which N_2O and HFC_{23} emissions are mitigated.

[10] See the UNFCCC website, accessible at http://cdm.unfccc.int/Issuance, accessed 4 April 2007. JI projects are eligible to generate and issue carbon credits only in the period 2008–12.

[11] According to article 3, paragraph 11, this is one of the objectives of the Kyoto Protocol.

[12] See the UNFCCC website 'Guide to do a CDM project activity', accessible at http://cdm.unfccc.int/Projects/pac, accessed 10 April 2007. More information on baseline methodologies can be found in a publication from the UNEP Risø Center: *Baseline Methodologies for Clean Development Mechanism Projects* (November 2005), available at www.cd4cdm.org. See also the CDM Guide from the Foundation Joint Implementation Network, available at: www.jiq.wiwo.nl, accessed 10 April 2007.

[13] See the UNFCCC website, accessible at www.cdm.unfccc.int/methodologies, accessed 10 April 2007.

[14] European Environment Agency, *Greenhouse Gas Emission Trend and Projections in Europe 2006* (Copenhagen, October 2006). Key GHG data – GHG emissions data for 1990–2003 submitted to the UNFCCC (November 2005).

[15]   World Bank, State and Trends of the Carbon market – 2006 (update Q3'06), Washington, DC, 2006.

[16]   The main reason for this difference is the smaller number of HF_{23} and N_2O projects in Latin America.

[17]   The CDM landfill gas methodologies are as follows (the models they refer to are in parentheses): AM0002 (first-order decay model), AM0003 (US EPA first-order decay model), AM0010 (US EPA first-order decay model), AM0011 (first-order kinetic model), and ACM0001 (verifiable methods for an *ex ante* estimate).

[18]   See the website of the Landfill Methane Outreach Program of the US EPA, accessible at www.epa.gov/landfill, accessed at 13 April 2007.

[19]   Scharff, H., ed. *Methods to Ascertain Methane Emission of Landfills*, (Assendelft: NV Afvalzorg, 2006).

[20]   See the IPCC website www.ipcc.ch, accessed April 2007.

[21]   K. Spookas, J. Bogner, J.P. Chanton et al. (2005), 'Methane Mass Balance at Three Landfill Sites: What Is the Efficiency of Capture by Gas Collection Systems?', *Waste Management*, 26(5), 516–25.

[22]   Findings based on experience of the authors and on pre-feasibility studies from SCS Engineers in Reston, Virginia. The studies were made for the World Bank in June 2005. They include the following landfills: El Carrasco landfill Bucaramanga, Colombia; El Combeima landfill in Ibague, Colombia; Muribeca landfill in Pernambuco, Brazil; Queretaro landfill in Queretaro, Mexico; Chihuahua landfill in Chihuahua, Mexico; Gramacho landfill Rio de Janeiro, Brazil; Huaycoloro landfill in Lima, Peru; and Montevideo landfill in Montevideo, Uruguay.

[23]   Current prices on the ETS, CDM and JI markets are well above the US$5 used in these estimates. However, publicly available information on the feasibility of projects is difficult to find; therefore, we choose to use this information from the World Bank.

[24]   See the World Bank website, accessible at: www.bancomundial.org.ar/lfg/gas_estudios_prefac_en.htm, accessed on 13 April 2007. The power sales price in Mexico was assumed to be 0.057 US$/kWh; in Brazil, 0.029 US$/kWh; in Uruguay, 0.027 US$/kWh; and in Peru and Colombia, 0.035 US$/kWh.

[25]   For a detailed explanation of CDM carbon contracts, see the CERSPA Guidance Document, available at www.cerspa.org.

[26]   Startup Journal –The Wall Street Journal Center for Entrepreneurs- 'Treaty Creates a Market For Gas-Emission Credits' (New York, 12 August 2005).

5

The economics of landfill mining and land reclamation

William Hogland

School of Pure and Applied Natural Sciences, University of Kalmar, Kalmar, Sweden; Department of Engineering, Physics and Mathematics, Mid Sweden University, Sweden

Kurian Joseph

Centre for Environmental Studies, Anna University, Chennai, India

Marcia Marques

Department of Engineering, Physics and Mathematics, Mid Sweden University, Sweden; Department of Sanitary and Environmental Engineering, Rio de Janeiro State University, Rio de Janeiro, Brazil

This chapter discusses a key dimension of 'WASTEnomics' – that of turning the liabilities of landfills into assets through landfill mining and land reclamation.

Land degradation

Land degradation due to anthropogenic effects is as old as civilisation. However, it was due to the Industrial Revolution, population growth and intensive urbanisation, as well as the 'green' (agriculture) revolution of the twentieth century, that anthropogenic land degradation reached its highest levels.

The global extent of land degradation due to different causes was estimated at 680 million ha due to overgrazing, 580 million ha due to deforestation, 550 million ha due to agricultural mismanagement, 137 million ha due to fuel-wood consumption, and 19.5 million ha due to industrialisation and urbanisation [1]. In the case of industrial and chemical pollution, because of the risk, these contaminated areas cannot be used for food production, raising animals, urban expansion or other economic purposes. A legacy of contaminated industrial and urban sites is common to all old industrial heartlands, particularly in the USA, Europe and the former USSR [2].

Across Europe, it is estimated that there may be more than 2 million such sites, containing hazardous substances such as heavy metals, cyanide, mineral oil and chlorinated hydrocarbons [3]. The number of contaminated sites estimated in different countries, including industrial, waste disposal and military sites, is 80,000 in Austria, 40,000 in Denmark, 300,000 in France, 240,000 in Germany, 110,000 in The Netherlands, and 50,000 in Switzerland, to mention only a few European countries [4]. Even in Sweden, a country that is among those with the highest environmental awareness and that has one of the best protection policies in the world, there are about 46,000 contaminated sites, and more than 10,000 of them belong to the highest risk categories and urgently need to be decontaminated [5].

A large number of contaminated sites have also been found in many countries with a transitional economy in Europe and in developing countries all over the world. In São Paulo State, for instance, where 55 per cent of the Brazilian industrial production is located, the environmental protection agency CETESB published a list of 2,272 contaminated sites in 2007 [6]. In developing countries, where the financial resources are limited, resource allocation usually does not prioritise remediation actions.

In the Baltic Sea drainage basin formed by 14 countries with about 85 million people [7], there are 70,000–100,000 old landfills, of which 10–20 per cent need remediation. The reason for this great number of landfills is that most of them are very small and of only local importance. In Sweden, 25 per cent of landfills are smaller than 500 m^3, 36 per cent are 500–2,000 m^3, 25 per cent are 2,000–20,000 m^3, and only 11 per cent are bigger than the last figure. This trend also applies to other countries in the region of the Baltic Sea. In Estonia, for instance, 57 per cent of landfills are less than 1.5 ha, 37 per cent are 1.5–5 ha, and only 6 per cent are larger than 6 ha [4]. The average landfill size in Japan is about 20,500 m^2 [8].

With its directive on waste landfills [9], the European Union (EU) intends to prevent or reduce the adverse effects of these landfills on the environment, and in particular on surface water, groundwater, soil, air and human health. It defines the different categories of waste (municipal waste, hazardous waste, non-hazardous waste and inert waste), and it applies to all landfills, defined as waste-disposal sites for the deposit of waste onto or into land. The directive is expected to change the situation in the EU for the future, resulting in increased incineration, and this tendency was clearly seen at the beginning of the new millennium. However, old landfills already closed or close to closure will cause emissions into air, soil and water if serious measures are not taken. Studies in the early 1990s in Sweden indicated that about 20–25 per cent of the landfill sites/dumps pose such a high environmental risk that they should be removed or remediated at once [4]. The environmental cost of the long-term emissions from these dumps is difficult to calculate.

Cities that expand as a result of rapid urbanisation have suddenly discovered an old dump, a contaminated industrial site, or a pond with oily waste and toxic chemicals in areas where business and shopping centres or multiple family dwellings are to be constructed. Clearly, there are economic purposes in reclamation of land, and these might include mining. In most cases, site remediation is intended to return the land to human use such as housing. However, restoration of a certain area might also create a new park in the city or restore a natural wildlife habitat. In several cases, town districts have been constructed on top of landfills that might contain toxic waste and emit toxic gases threatening human health.

Landfill mining

Landfill mining is the process of excavating, sorting, recycling, processing or sending for other dispositions fractions of solid waste that had been previously landfilled [10–15]. A typical landfill-mining scheme is presented in Figure 1.

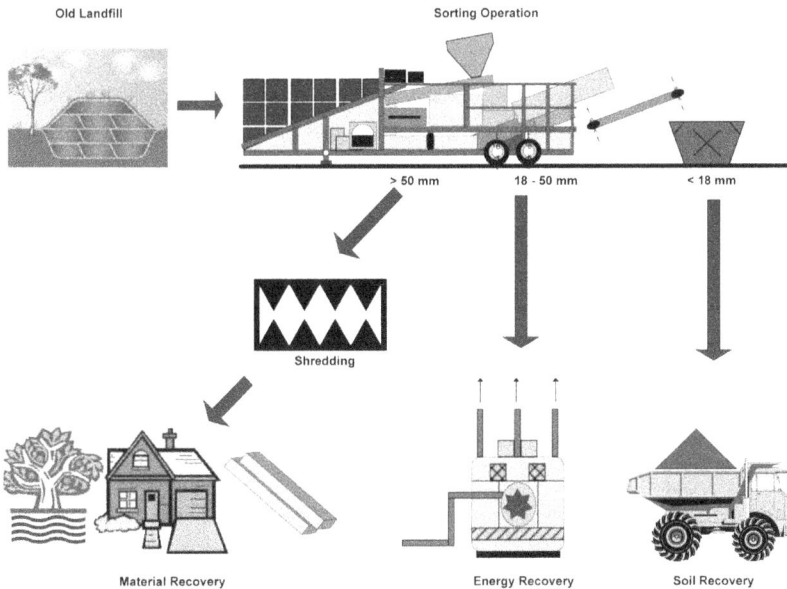

Figure 1. Schematic of a landfill mining process

Landfill mining employs the method of open-cast mining to sort out with a screening machine the mixed material from the landfill by size. The key to landfill mining is a set of conveyers and screens that sort out the solid waste into three sizes: oversized, intermediate-sized, and dirt/humus [16]. The oversized materials are pre-screened by another sorting machine that separates the larger objects, such as tyres and rocks, from cardboard and other smaller, unearthed materials.

Landfill mining typically involves excavation and a series of mechanical operations to recover part of or all space occupied by a landfill. Additionally, it allows recovery of compost/soil that can be used on site as daily cover material for new landfill cells; metals such as iron, aluminum and copper; plastics; wood; glass; concrete; bricks;

stones; and mortar material for road construction. The success of material recovery depends on the composition of the waste, and the effectiveness and efficiency of the mining technology [11]. If necessary, air can be pumped into the landfill before excavation in order to interrupt the anaerobic decomposition and methane production, and thus reduce the risk of explosion.

Combustible waste reclaimed by mining can be mixed with fresh waste and burned for energy utilisation. By reducing the size of the landfill's 'footprint' through cell reclamation, the operator may be able either to lower the cost of closing the landfill or make land available for other uses and even for proper construction of a new sanitary landfill. Hazardous wastes, especially in older landfills, can be managed in an environmentally sound manner in insolated cells or destroyed in special incineration plants.

Landfill mining limitations and constraints

Small, shallow landfills might be easier to excavate from a technical point of view, but when the size of the landfill increases surface area and depth, it must reach a certain scale before development becomes economically beneficial, as the capital cost might be high. It is necessary to collect the site-specific information and site characteristics before starting to plan the landfill-reclamation project, and the assessment of potential economic benefits also includes a survey of the markets for recovered materials.

One limitation of landfill mining is that it requires an intensive use of machinery and manpower. Other limitations include odour and air emissions at the reclamation site, increased traffic on roads between the landfill and the resource-recovery facility, extra mixing and handling of waste at the resource recovery facility, and handling of additional inert materials. Reclamation activities shorten the lifetime of equipment, such as excavators and loaders, as the high particulate content and abrasive nature of reclaimed waste increase mechanical wear.

Lack of knowledge of the nature of buried waste might be a limitation with respect to safety; the hazards include physical injury from rolling stock or rotating equipment, exposure to leachate and hazardous material or pathogens, subsurface fires, and landfill gas emissions. Excavation of one landfill area can undermine the integrity of adjacent cells, leading to sinking or collapse into the excavated area. Health risks to the general public appear to be minimal. During rainy periods, the generation of storm water can be large and the environmental effects considerable. This might place extra load on an existing leachate-treatment system.

There is considerable concern about the working environment due to the burial of hazardous materials in many landfills and the presence of explosive gases such as methane. Modern safety equipment and precautionary measures are needed during landfill mining. This may include safety goggles, hard hats, respirators, first-aid kits, leather work gloves, hearing protection, back supports, steel-toed work boots, combustible gas meters, oxygen analysers, hydrogen sulfide chemical reagent diffusion tube indicators, and a water spray system to suppress dust.

Case studies and lessons learned

Landfill mining has been applied throughout the world during the last 50 years as a method of sustainable landfilling [17, 18]. The first reported landfill mining project was an operation in Tel Aviv, Israel, in 1953, which was then done to recover the soil fraction and improve the soil quality in orchards [19, 20]. This procedure was later employed in the USA to obtain fuel for incineration and energy recovery [11, 21, 22]. Pilot studies carried out in the UK, Italy, Sweden, Germany [23, 24], China and India have been reported. Landfill mining has been reported as a method of waste management in many developed and some developing countries [25–28]. Cossu et al. [11] reported on the technical and practical experience gained on several commercial landfill-mining projects in the USA and pilot and research experience in Europe. In 1995, a manual for landfill reclamation was published [29].

The recovery of various materials ranges from 50 to 90 per cent of the waste [15]. The average soil fraction in recovered municipal waste from landfill tends to be around 50–60 per cent. However, it varied between 20 and 80 per cent [4] when studies from 15 excavated landfills in four countries in two continents were compared, and the soil fraction rate depended on moisture content and decomposition rate as well as on the age of the landfill. Savage et al. [20] reported the following proportions of various materials that were recovered: soil, 85–90 per cent; ferrous metals, 80–95 per cent; and plastic, 70–90 per cent.

Case study: Filborna, Sweden

During the summer of 1994, a 10-year-old part of the Filborna landfill in Sweden was excavated as a pilot test in a research project [24]. The waste disposed consisted of a mixture of household, industrial, construction and demolition waste. About 1,300 m³ of waste was excavated to a depth of 8.5 m. The excavation was made in two stages: first to the 5-m level, and then to 8.5 m over a plot size of 30 m². The characteristics of the material obtained from the landfill-mining studies are shown in Tables 1 and 2.

Table 1. Characteristics of the mined waste

Level below surface	pH	Temp. °C	CH$_4$ %	CO$_2$ %	Coarse fraction: amount by volume, amount by weight, density and moisture				Fine fraction: amount by volume, amount by weight, density and moisture			
					By vol. %	By wt. %	Dens. t/m³	Moist. by wt. %	By vol. %	By wt. %	Dens. t/m³	Moist. by wt.
0–5 m	4–5	17	–	–	35	45	0.5	38	65	55	0.4	30
5–8 m	6.5	18–20	59	40	70	25	0.4	43	30	70	2.5	39

Source: Hogland et al. [24].

Table 2. Total solids, ash content, low calorific value and concentration of different constituents in the waste at 0–5 and 5–8 m below the surface

Parameter	Unit	Coarse fraction 0–5 m	Fine fraction 0–5 m	Coarse fraction 5–8 m	Fine fraction 5–8 m
Total solids (TS)	(%)	62.0	70.0	56.6	61.0
Ash content	% of TS	39.3	78.9	36.6	84.0
Calorific value	MJ/kg sample	6.9	<2	7.9	<2
Carbon (C)	% by weight TS*	32	13	44	11
Nitrogen (N)	% by weight TS*	0.74	0.45	0.49	0.57
Sulphur (S)	% by weight TS*	0.39	0.71	0.27	0.56
Phosphorus (P) (tot)	g/kg TS*	0.77	0.72	0.66	1.5
COD$_{Cr}$	g/kg TS*	720	250	620	270
Magnesium (Mg)	g/kg TS*	0.84	1.6	0.99	1.6
Calcium (Ca)	g/kg TS*	12	17	7.6	15
Potassium (K)	g/kg TS*	1.4	0.99	0.85	1.3
Zinc (Zn)	g/kg TS*	1.9	0.50	0.33	0.58
Nickel (Ni)	mg/kg TS**	6.7	12	8.7	30
Copper (Cu)	mg/kg TS**	90	53	41	140
Chromium (Cr)	mg/kg TS**	0.39	36	8.1	39
Lead (Pb)	mg/kg TS**	88	160	18	100
Cadmium (Cd)	mg/kg TS**	7.1	1.6	0.57	3.4

*Calculated based on the whole sample

**Calculated based on the whole sample, but for the fractions metals, glass, stone, etc.

Source: Hogland et al. [24].

Case study: Måsalycke, Sweden

From a test excavation at Måsalycke landfill in southern Sweden, Hogland [30] reported on the composition of different fractions of material after a screening into three sizes (Table 3). The medium-sized fraction, 18–50 mm, exhibited a much higher content of stones and waste than the others. The organic content was substantial. Treatment options for this fraction may be digestion/methane gas fermentation or combustion. Some metals could also be recovered. The coarse fraction, of >50 mm, contained large amounts of paper (29 per cent) and wood (19 per cent), and the calorific value was high if the material was dry. The metal content (5 per cent) was higher than in the other fractions, and metal recovery may be possible. The dry-matter concentration found at Måsalycke landfill was 70–80 per cent, while the ash content was about 90–95 per cent in the two finest fractions. The calorific value was very low in those two fractions: 0–1 MJ/kg. Therefore, the fine fraction was unsuitable for further biological or thermal treatment. The fraction of >50 mm consisted mainly of paper, wood and plastic and therefore had a higher calorific value, 7 MJ/kg or more.

In Sweden, the soil fraction is considered the most valuable fraction; thus, it can be used as final covering material for landfills. In some areas, there is lack of such material, and it must be bought at a relatively high price and sometimes transported over long distances.

Table 3. Composition of samples of the Måsalycke landfill [30]

Fraction size (%)	Unsorted	<18 mm	18–50 mm	> 50 mm
Paper	9.7	0.8	11.2	28.7
Plastic	4.9	0.2	3.7	6.5
Nappies, sanitary towels	0.5	–	–	2.1
Textiles	2.3	0.0	0.3	1.2
Rubber	0.6	0.0	0.0	0.6
Leather	0.0	0.0	0.0	0.2
Glass, ceramics	0.3	1.9	2.6	0.1
Metal	1.7	0.1	2.2	4.9
Food waste	0.5	0.0	0.8	1.2
Electronics	0.0	0.0	0.1	0.1
Garden waste	1.1	0.0	6.7	6.2
Wood	9.9	0.4	10.2	18.6
Stones, etc.	13.7	2.4	31.6	10.3
Hazardous waste	0.2	–	0.4	2.4
Indefinable*	54.5	94.1	30.2	16.9

*Fine fraction.

Cossu et al. [23] found the energy value of excavated waste in Italy to be 3.4–8.7 MJ/kg (mean 4.5 MJ/kg). The corresponding value for waste excavated from the Filborna landfill in Sweden was 6.9–7.9 MJ/kg in the light fraction and less than 2 MJ/kg in the fine fraction. Obermeier and Saure [31] obtained a value of 11 MJ/kg, and Cossu et al. [23], Rettenberger [32] and Schillinger et al. [33] found values up to 20 MJ/kg in the unsorted light fraction.

These studies support the feasibility of recovering about 50 per cent of the material at the dump site as valuable compost that can be used as daily cover for the landfill or, off-site, for growing non-edible plants. Based on the results of these studies, the following recommendations can be made for the reclamation operations at the landfill and resource-recovery facility: (1) proper planning of the excavation site to control the flow of storm water and methane; (2) reliable methods for measuring volumes and tonnes of reclaimed waste, cover soil, and non-combustible material and to track volumes by field-survey methods; (3) daily observation of the reclaimed waste and proper records of moisture content, waste composition, waste age, soil content of refuse, rainfall, weather, and odour; (4) minimising personnel exposure to the actual reclamation site during trammeling operations; (5) optimum mix of municipal solid waste (MSW) and reclaimed waste to maximise combustion efficiency; (6) daily monitoring of methane, oxygen, and volatile organic compounds (VOCs), and establishing action levels for each parameter; (7) quarterly physical and chemical characterisation of screened wastes.

Economic aspects

Most potential economic benefits associated with landfill mining may include any or all of the following: (1) the land value of sites reclaimed for other uses; (2) increased disposal capacity if the landfill site remains active; (3) avoided or reduced costs of landfill closure and post-closure care and monitoring; (4) revenues from recyclable and reusable materials, such as ferrous metals, aluminium, plastics, and glass. The major benefit of the mining approach is certainly related to the recovery of land for use. An extension of many years of the lifetime of the existing landfills is usually the purpose, obviating the cost and time needed to locate, design, permit and construct a new landfill or implement alternative waste management.

Major factors influencing the cost of landfill reclamation projects include the volume and topography of the site, equipment parameters, soil conditions, climate, labour costs, the regulatory approval process, excavation and screening costs, sampling and characterisation, development costs, the contractor's fees, hazardous waste disposal, and revenue from the sale of commodities such as compost and recyclables. The cost is expected to decline as the industry gains experience and new equipment and techniques become available. Gradually, landfill mining will become more attractive to towns facing dump/landfill closure or operators wanting to extend the life of their landfills.

Expenses incurred in project planning, including capital and operational costs of a landfill mining project, include the following [34]:

- *capital costs*: site preparation, rental or purchase of reclamation equipment, rental or purchase of personnel safety equipment, construction or expansion of materials-handling facilities, rental or purchase of hauling equipment;

- *operational costs*: labour (e.g. equipment operation and materials handling), equipment, fuel and maintenance, administrative and regulatory compliance expenses (e.g. record keeping), worker training in safety procedures, hauling costs.

As mentioned previously, the greatest potential economic benefits associated with landfill reclamation are indirect. However, a project can generate revenues if markets exist for recovered materials. Although the economic benefits of reclamation projects are facility-specific, they may include any or all of the following [34]:

- increased disposal capacity;

- avoided or reduced costs of landfill closure, post-closure care and monitoring, purchase of additional capacity or sophisticated systems, and liability for remediation of surrounding areas;

- revenue from recyclable and reusable materials (e.g. ferrous metals, aluminium, plastic, and glass), combustible waste sold as fuel, reclaimed soil used as cover, materials sold as construction fill or sold for other uses, and land value of sites reclaimed for other uses.

Selected studies: the USA and Canada

A comprehensive field test evaluation of the Collier County landfill mining system was conducted in 1992 under the Municipal Innovation Technology Evaluation (MITE) Program of the US Environmental Protection Agency [34]. The mining operation reclaimed 50,000 tonnes of soil suitable for use as landfill cover. By 1995 prices, the reclaimed cover soil had a cost saving of US$1 per tonne compared to conventional cover.

In 1986, the town of Thompson, Connecticut, initiated a landfill-mining project with the objective of recapturing landfill volume and extending the life of the landfill temporarily [15]. At the time of the mining project, the available disposal alternatives represented costs in the range of US$66–88/tonne, including transportation. The cost of the mining project was US$117,000 including grading the base of the mined area to receive new MSW. Representatives from the town estimated that the town saved US$1 million in tipping fees over an 18-month period.

The first effort in the USA to dig up and entirely remove an old landfill to return the site to its natural state was the Hague Landfill Reclamation Project [26]. The project aimed to remove a 2.7-ha landfill from the middle of a 52-ha site owned by the rural township for the purpose of using the land for recreational purposes. About 76,500 m³ was removed and separated for recovery of ferrous metal and for the beneficial use of soil fraction. The project budget was US$1.3 million. Based on the information developed by Landfill Mining, Inc., from its operation in the Collier County at 1995 prices, the cost of landfill mining is expected to be less than about US$10/tonne of waste mined. A large amount of that cost is associated with rental of the processing equipment. The rental fee is typically US$16,000–19,000 per month. For a large-scale plant in Europe, a cost of US$75–100/m³ was reported [11].

The results of an analysis of the weekly production data, project costs and assets realised during 1992 and 1993 at the Frey Farm Landfill of Lancaster County, Pennsylvania, showed that 33 per cent of the project costs were associated with excavation and trammeling operations at the landfill. Transportation of reclaimed waste to the resource-recovery facility (RRF) and hauling ash residue back to the landfill incurred 30 per cent of the cost. The balance of the project costs was associated with processing fees paid to the landfill mining operator, RRF and landfill host communities. Revenues obtained from the sale of electricity from the RRF and recovered ferrous metal offset these operating costs and resulted in net revenues of US$3.94 for every tonne of reclaimed material delivered to RRF. Additional assets recovered included cover soil and landfill volume, making the overall profit US$13.30 for every tonne of material excavated.

The McDougal project (Ontario, Canada), started in 1994, aimed to remove the entire 3-ha landfill cell, line the site, and put the waste back in after screening with a power screen trammel to remove soil fraction [26]. The project was undertaken to remediate leachate problems at the landfill after contaminants were found in monitored wells. In addition, the project was expected to have enhanced the landfill capacity by 5–10 years. About 50 per cent of the reclaimed waste was soil, most of which was used as daily cover and landscaping. The total budget, including relining, was US$7 million.

Selected studies: Europe and Japan

The main purpose of the first landfill mining in Europe, at the Burghof landfill site in 1993 [22, 32], was environmental remediation and the construction of new landfills by modern technology. A total of 53,700 tonnes of material was excavated and sorted from the landfill in 14 months. About 70.5 per cent of the reclaimed waste by weight was fine fraction and was reused at the landfill. About 17.5 per cent of the reclaimed waste was light fraction and was used at a waste-to-energy facility. The project helped achieve additional volume for waste deposition, improve the long-term behaviour of the displaced waste, assess the technical and economical feasibility of landfill mining, and define more suitable measures to ensure optimal environmental conditions for workers and the neighbourhood [11]. The test screening and the recovery of material from the Måsalycke landfill, as well as various other projects, showed that excavation is a realistic alternative to lifetime expansion and remediation of small and medium-size landfills [4]. The cost of landfill mining at the Filborna landfill in Sweden in 1994 was US$6.7/tonne.

The Japan Environmental Sanitation Center carried out a small-scale feasibility study of an old landfill, 25,000 m^2 in size and 200,000 tonnes in weight, including melting of excavated materials. The estimated capital cost for excavation and separation was US$10–30 per tonne waste, and the operation cost was US$30–90 per tonne waste. In total, this was US$40–120 per tonne, and the melting cost totalled US$325–400/tonne [35].

Conclusion

Landfill mining as a method of waste management is yet to be widely tested at the commercial level. The quantity and characteristics of materials recovered from a landfill are dependent on the disposed waste, the selected mining technology, and the efficiency with which the technology is applied. Based on the studies reviewed here, the characteristics of the heavy metal content and soil fractions indicate that the fraction could be suitable for landfill cover. However, it is possible that high concentrations of hazardous substances and heavy metal could be found in local pockets in the landfills.

The costs and benefits of landfill mining vary considerably depending on the objectives (closure, remediation, new landfill, etc.) of the project, site-specific landfill characteristics (material disposed, waste decomposition, burial practices, age and depth of fill), and regional economics (value of land, cost of closure materials and monitoring).

In general, the economics of landfill mining depends on the depth of the waste material and the ratio of wastes to soil. In most cases, the presence of hazardous materials and legal requirements for proper handling will also affect the economic feasibility. Analysing the economics of landfill mining calls for investigating the current landfill capacity and projected demand, the projected costs of landfill closure or expansion of the site, the current and projected costs of future liabilities, the projected markets for recycled and recovered materials, and the projected value of land reclaimed for other uses. The size and the volume might have an important influence on the suitability of reclamation.

Acknowledgements

The Kalmar Research and Development Foundation and Graninge Foundation are acknowledged for their sponsorship. The support of the waste management companies NSR and ÖKRAB is also acknowledged. The authors would like to thank the Swedish International Development Cooperation Agency (Sida) for the important financial contribution to the subject through the Sustainable Solid Waste Landfill Management in Asia under the Asian Regional Research Programme on Environmental Technology being coordinated by Asian Institute of Technology, Bangkok, Thailand.

References

[1]   UNEP. *GEO 3: Global Environmental Outlook: United Nations Environmental Programme*, 2002 (www.unep.org/GEO/geo3/), accessed December 2007.

[2]   UNEP. *GEO 4: Global Environmental Outlook: United Nations Environmental Programme*, 2007 (www.unep.org/geo/geo4/media/index.asp), accessed December 2007.

[3]   EEA. *The European Environment – State and Outlook 2005* (Copenhagen: European Environment Agency, 2005).

[4]   Hogland, W. (2002) 'Remediation of an Old Landfill Site – Soil Analysis, Leachate Quality and Gas Production', *ESPR – Environmental Science and Pollution Research* (Special Issue), 2002, 1, 49–54.

[5]   Swedish Environmental Objectives Council. *Sweden's Environmental Objectives*, 2005.

[6]   CETESB. *O Gerenciamento de áreas contaminadas no Estado de São Paulo*, 2007 (www.cetesb.sp.gov.br/Solo/areas_contaminadas/relacao_areas.asp), accessed December 2007.

[7]   Gren, I.-M., Turner, R. and Wulff, K., eds. *Managing a Sea. The Ecological Economics of the Baltic* (London: Earthscan Publications, 2000).

[8]   LSA. *Landfill Sites in Japan 2000* (Tokyo: Landfill System Technologies Research Association, 2000).

[9]   EU. EU Directive on Landfill of Waste. European Union Council Directive 1999/31/EC issued 26 April 1999.

[10]   Lee, G.F. and Jones, R.A. 'Use of Landfill Mining in Solid Waste Management' in *Proceedings of the Water Quality Management of Landfills Conference*, Chicago: Water Pollution Control Federation, 1990), p.9.

[11]   Cossu, R., Hogland, W. and Salerni, E. *Landfill Mining in Europe and the USA* (ISWA Year Book, International Solid Waste Association, ed., 1996), pp.107–14.

[12]   Hogland, W. and Marques, M. 'Landfill Mining – Space Saving, Material Recovery and Energy Use' in *Proceedings of the Ecological Symposium, 5–7 June 1998* (Gdańsk: Oruma, 1998), p.12.

[13]   Hogland, W. and Nimmermark, S. 'Assessment of Waste Masses in Old Landfills in Sweden' in *Proceedings of the 2nd International Youth Environmental Forum ECOBALTIC '98, 22–26 June 1998, St Petersburg.*

[14]   Carius, S., Hogland, W., Jilkén, L., Mathiasson, A. and Andersson, P.-Å. (1999) 'A Hidden Waste Material Resource: Disposed Thermoplastic' in *Proceedings of Sardinia '99, the 7th International Waste Management and Landfill Symposium, 4–8 October 1999, Cagliari, Italy*, pp.229–35.

[15]   Strange, K. *Landfill Mining* (Tonbridge: World Resource Foundation, 1998) (www.cbvcp. com/columbiasd/techpage), accessed May 2003.

[16]   Hogland, W., Marques, M. and Nimmermark, S. 'Landfill Mining and Waste Characterization: A Strategy for Remediation of Contaminated Areas', *Journal of Material Cycles and Waste Management*, 2004, 6(1), 119–24.

[17]   Joseph, K., Esakku, S., Palanivelu, K. and Selvam, A. 'Studies on Landfill Mining at Solid Waste Dumpsites in India' in *Proceedings of Sardinia '03, the 9th International Waste Management and Landfill Symposium, 6–10 October 2003, Cagliari, Italy.*

[18]   Joseph, K., Nagendran, R., Palanivelu, K., Thanasekaran, K. and Visvanathan, C. *Dumpsite Rehabilitation and Landfill Mining* (Chennai, India: CES, Anna University, ARRPET, 2004).

[19]   Shual, M. and Hillel. 'Composting Municipal Garbage in Israel', *Tavruau*, 1958, July–December.

[20]   Savage, M.G., Golueke, C.G. and Stein, E.L. 'Landfill Mining – Past and Present', *Biocycle*, 1993, May.

[21]   Hogland, W., ed. *Workshop on Landfill Mining* (Piteå Havsbad, Sweden: Swedish Research Council, 1996).

[22]   Hogland, W., Marques, M. and Thörneby, L. (1997) 'Landfill Mining – Space Saving, Material Recovery and Energy Use', in *Proceedings of Seminar on Waste Management and the Environment – Establishment of Cooperation Between Nordic Countries and Countries in the Baltic Sea Region, 5–7 November 1997, Kalmar University, Kalmar, Sweden*, pp.339–55.

[23]   Cossu, R., Motzo, G.M. and Laudadio, M. (1995) 'Preliminary Study for a Landfill Mining Project in Sardinia', in *Proceedings of Sardinia '95, the 5th International Landfill Symposium, 2–6 October 1995, Cagliari, Italy*, vol. 3, pp.841–50.

[24]   Hogland, K.H.W., Jagodzinski, K. and Meijer, J.E. (1995) 'Landfill Mining Tests in Sweden' in *Proceedings of Sardinia '95, the 5th International Landfill Symposium, 2–6 October 1995, Cagliari, Italy.*

[25]   Murphy, R.J. *Optimization of Landfill Mining* (Tampa, FL: Department of Civil Engineering and Mechanics, University of South Florida, 1993).

[26]   Nelson, H. 'Landfill Reclamation Projects on the Rise', *Biocycle*, 1995, March, pp.83–4.

[27]   Foster, G.A. 'Assessment of Landfill Reclamation and the Effects of Age on the Combustion of Recovered MSW', Municipal Solid Waste Management, Forester Communications, Inc., 2001, March/April.

[28]   Hull, R.M., Krogmann, U. and Storm, P.F. 'Characterization of Municipal Solid Waste Reclaimed from a Landfill' in *Proceedings of Sardinia 2001, the 8th International Waste Management and Landfill Symposium, Cagliari, Italy, 2–6 October 2001*, pp.567–76.

[29]   Salerni, E.L. 'Landfill Reclamation Manual' in *Reclaim '95 Landfill Mining Conference, 28–29 September 1995*, SWANA Landfill Reclamation Task Group.

[30]   Hogland, W. (2001) 'Landfill Mining and Remediation of Soils' in *Book of Abstracts, Workshop on Management of Industrial Toxic Wastes and Substances Research – Bioremediation and Polluted Ecosystems, the 2nd CCMS/NATO Workshop, Instituto Le Monacelle, Matera, Italy, 20 December 2001*.

[31]   Obermeier, T. and Saure, T. 'Landfill Reconstruction, Biological Treatment of Landfill Waste' in Christensen et al., ed., *Proceedings of Sardinia '95, the 5th International Landfill Symposium, Cagliari, Italy, 2–6 October 1995*, vol. 3, pp.819–26.

[32]   Rettenberger, G., Urban-Kiss, S., Schneider, R. and Goschl, R. 'German Project Reconverts a Sanitary Landfill', *Biocycle*, 1995, June, pp.44–8.

[33]   Schillinger et al. 'Summary of Landfill Reclamation Feasibility Studies', NYSERDA, 1994.

[34]   USEPA (1997) 'Landfill Reclamation', United States Environmental Protection Agency, Solid Waste and Emergency Response (5306W), EPA530-F-97-001, July 1997.

[35]   Inanc, B., Yamada, M., Ishigaki, T. et al. 'The Need for Landfill Reclamation in Japan', International Landfill Research Symposium, 2002.

[36]   ARRPET (Asian Regional Research Programme on Environmental Technology). Swedish International Development Cooperation Agency (www.arrpet.ait.ac.th/), accessed November 2007.

Further References

Esakku, S., Palanivelu, K. and Joseph, K. (2003) 'Assessment of Heavy Metals in a Municipal Solid Waste Dumpsite' in *Proceedings of the Workshop on Sustainable Landfill Management, Chennai, India, 3–5 December 2003*, pp.139–45.

Joseph, K., Esakku, S. and Nagendran, R. 'Mining of Compost from Dumpsites and Bioreactor Landfills', *International Journal of Environmental Technology and Management*, 2007, 7(3–4), 317–25.

Savage, G.M., Diaz, L.F. and CalRecovery, Inc. 'Landfill Mining and Reclamation', *ISWA Times*, 1994, 4, 1–4.

6
Redefining the changing world of waste projects – turning waste into energy

Paul Carey
Waste Recycling Group plc, UK

This chapter discusses a key dimension of 'WASTEnomics' – that of turning the liabilities of municipal solid waste into an asset and a valuable resource of power through the use of energy from waste (EfW) plants.

Introduction

Over the last decade and a half, the world of waste management has been changing. The pace of that change has been remarkable, and is set to accelerate rapidly as we approach the end of the first decade of the new century.

Increasing environmental awareness in government and among the public at large, driven by pressure from the European Union (EU), and organisations such as Greenpeace and Friends of the Earth, has prompted a wholesale change in the way we now view our waste. EU legislation has, in particular, forced a rapid change in the way waste is perceived and dealt with. More than 70 sets of laws and regulations have been implemented in the UK since and including the introduction of the Environmental Protection Act 1990, dealing with issues such as competitive tendering of local authority contracts, recycling levels, standard of emissions from incinerators, and, most recently, diversion of biodegradable waste from landfills.

Under this changing regulatory structure, waste has gone from being just that – a waste product to be disposed of – to being seen as a valuable resource from which value can be created.

In response, the waste-management industry has had to change. Long gone are the days of a regional and local industry filling quarries and other holes in the ground with rubbish. Today, the industry is dominated by a small number of major players, most of which have an international parentage. These companies can achieve economies of scale to enable them to offer competitively priced services of a complex nature to local authorities, who now have expert procurement and service supervision capabilities.

The technologies which support the delivery of those services have also changed. Landfills are now highly engineered structures with impermeable linings, leachate drainage and treatment systems, and capping and gas-collection systems with power generation, and they are subject to intense environmental scrutiny and liabilities on the part of the operator. But still this is not enough. Under the EU Landfill Directive, the UK has implemented a system of strict limits on the amount of biodegradable waste that is landfilled, with stiff penalties for waste-disposal authorities that exceed those limits. This legislation, together with clear targets for recycling waste, is leading to a wholesale change in the methods by which we deal with our rubbish. Nowadays, industry jargon is full of acronyms such as MBT, EfW and ATT (more of which will be explained later), not to mention the plethora of acronyms surrounding the procurement of waste-management services.

The unpalatable truth is also that the cost of waste management has increased and is still increasing. Landfill gate fees have increased though the incorporation of not just improved technical standards but also legislated requirements for aftercare, the landfill tax, and, most recently, landfill allowances. Energy-from-waste (EfW) gate fees have risen alongside those of landfill. The cost of the disposal of residual waste in an EfW facility has gone from less than £20/tonne in 1991 to almost four times that for a similar-size plant.

Yet, despite the advances in waste management since 1990, the debate over the manner and method in which our rubbish is dealt with is as intense as ever, and thus the way forward for local authorities is often not clear. What is clear though is that there is no single solution that fits all circumstances. It is also clear that investments in many 'new' technologies are unlikely to bear fruit, with just a handful likely to come to the fore as proven and (possibly) acceptable ways of dealing with our rubbish. What is also clear is that when we view it as a resource, there is an increasing overlap between the waste and the energy sectors, in particular power (as in electricity generation).

Moreover, planning policy is increasingly placing a strong emphasis on the recovery of energy from waste, particularly in the form of combined heat and power.

Box 1. Waste-management regulation

The underlying principle which now drives the management of waste is the waste-management hierarchy:

> Reduce
> Reuse
> Recycle
> Recover
> Dispose

To achieve these elements, the waste industry is now governed by a number of specific measures, given force through legislation.

Packaging regulations

The packaging regulations were introduced in 2005 and are aimed at packaging manufacturers with over £2 million turnover or that handle over 50 tonnes of packaging. The aim is to promote the recycling and recovery of packaging, and to reduce the amount used. The target is for 60 per cent of all packaging to be recycled or recovered by 31 December 2008.

This is being achieved through the buying and selling of Packaging Waste Recovery Notes (PRNs), which represent the recycling or recovery of an equivalent weight of the material used within the packaging. PRNs are sold by any company recycling or recovering material.

Recycling targets imposed on waste-collection and waste-disposal authorities

Recycling targets are a statutory requirement imposed on waste-collection authorities by central government. These are then used as an annual measure to assess performance by local authorities. These targets are also used as a gauge to show how likely an authority is to achieve its landfill diversion targets in 2010, 2015 and 2020.

The government has set what it believes are achievable levels of recycling tailored to each authority. For example, the 2005/6 target for Hampshire was 30 per cent; for Tower Hamlets in London, it was 18 per cent.

Landfill tax

The various landfill regulations impose a tax on each tonne of waste sent to landfill. Present levels range from £2/tonne for inert waste (e.g. construction-demolition waste such as rubble, or ash from waste plants) to £21/tonne for 'active' waste, that is, waste that is biodegradable or chemically reactive. The latter figure will rise by £3/tonne every year until it reaches a cap of £35/tonne in 2012. As mentioned earlier, landfills today are advanced structures, with methane-collection and power-generation systems. The total installed UK landfill gas power capacity is now about 600 MW, about 1.6 per cent of the UK's total demand. The landfill tax leads to much higher landfill costs than seen before 1990. On top of commercial gate fees for landfill of £10–£30/tonne, the avoided cost of landfill was often still too low to enable other technologies to be competitive. Therefore, diversion of biodegradable waste from landfills was introduced.

Diversion of biodegradable waste

The EU landfill directive was implemented in UK law through various regulations in 2004 and 2005. For example, under these regulations, English waste-disposal authorities will pay £150/tonne of biodegradable waste (deemed by the government to be 68 per cent of household waste) sent to landfill in excess of set targets expressed as a percentage of 1995 landfill tonnages (Table 1).

Table 1. Diversion targets

Year	%
2010	75
2013	50
2020	33

The English Landfill Allowances Trading Scheme enables authorities that achieve better than their targets to sell their surplus allowances to other authorities who have failed to do so. Trading commenced in 2005, and the initial market value was understood to be about £30/tonne, although trading volumes were predictably thin as the targets have so far been relatively easy to beat. The critical years for 2010 and 2013 are forcing many authorities to bring forward the procurement of long-term waste-management service contracts. These often involve the provision of complex multi-element services, including management of household waste-recycling centres (community tips), disposal of kerbside collected dry recyclables, separation and recovery of further recyclables from residual waste, disposal or composting of green waste, and, last but not least, disposal of the residual fraction left after all the other activities have been carried out. It is this last element that has excited the most controversy within the waste-management sector. It is also where there are the most opportunities for overlap with the power sector.

Industry estimates vary widely but the amount of residual waste which will need to be dealt with other than by landfill in 2020 is around 20 million tonnes per annum. Some formal reports have suggested that there will be sufficient waste to provide up to 17 per cent of the UK's power demand [1].

So what technologies are available to enable this diversion objective to be achieved? The answer is many that are being developed with a view to the future, but not so many that are proven now and can be implemented in projects that are funded by the most common form of finance – limited recourse project finance.

Many technologies are at a stage of development that makes them unsuitable for the larger projects required by many local authorities. Many simply do not work, or are of a kind where it is difficult to see what added value they bring in terms of diverting biodegradable waste from landfill. Some are simply pretreatment technologies which (in theory at least) make it easier to separate the waste into its component fractions, so that recovery and recycling of useful elements (e.g. certain types of plastic) are easier. Some are ingenious 'black boxes' that can achieve wonderful things that the promoters claim will save the planet, but are in fact unproven.

Those that are in the usable, bankable category, and realistically capable of helping authorities (through their selected service providers) to achieve their diversion targets are as follows:

- anaerobic digestion (AD);

- mechanical biological treatment (MBT) with a 'compost' output;

- MBT with a refuse-derived fuel output;

- energy from waste (EfW).

Let us deal with each of these in turn.

Anaerobic digestion (AD)

AD is essentially the controlled acceleration of the natural biodegradation of the organic component of the waste. This fraction is either collected separately (kitchen waste) or can be separated out in a 'dirty MRF' (materials-recycling facility) from the residual waste by means of a trommel screen. In both, the organic waste is kept in a warm tank in a wet sludge or liquid where natural bacteria decompose the waste. Different bacteria operate at different temperatures, but the most common is the thermophilic range of about 55 °C.

Methane is the main gas produced (as opposed to carbon dioxide in the case of aerobic digestion), but other gases, such as hydrogen sulphide, are also produced, resulting in odours if not properly controlled.

While the actual process is a natural one, the techniques of doing this in a cost-effective manner vary. There are a number of proprietary (including patented) systems, many of which are bankable.

The big advantage of AD is that it is apparently without emissions. That is true for the AD process itself, but emissions still occur when the gas is used as a fuel. If used as a fuel for power generation, the electricity can earn Renewable Obligation Certificates (ROCs) in the UK, thus earning almost three times the normal cost of power.

If not used as a fuel, the gas must be flared, as it is some 20 times more potent as a greenhouse gas than the carbon dioxide that would otherwise be produced. Thus, increasingly, consideration is also being given to the use of the gas as a compressed natural gas road fuel, often for municipal vehicles (refuse collection lorries, buses, etc). A challenge indeed!

Mechanical biological treatment (MBT) with a 'compost' output

MBT is a generic term that covers a wide range of systems, some proprietary technologies, and also a loose assembly of different manufacturers' equipment. Either way, MBT falls into two basic categories; those which produce 'compost' with no energy recovery, and those which produce refuse-derived fuel (RDF), sometimes also called solid recovered fuel (SRF – see later section).

The compost produced is not compost in the sense of something a domestic gardener would use. Rather, it is a material that introduces selected limited organic components into the ground – often termed 'soil conditioner'. Such compost is mostly used as a capping material in landfills or for final restoration (to avoid the use of excavated material), and it often contains residual traces of plastic and other compounds.

This form of MBT uses much of the innate energy in the waste to drive off moisture, but, instead of generating energy, it consumes it. When one looks closer at the technologies employed to reduce the biodegradable content of the waste (which, let us remind ourselves, is what all the effort is about), this form of MBT could be likened to above-ground landfill! In many senses, it is 'waste of energy', and, as such, does not deserve further mention in this chapter.

MBT with a Refused Derived Fuel (RDF) output

Several proprietary MBT systems openly use some of the organic energy in the waste, together with energy in the form of power from the grid, to dry the waste. In theory, this at least makes it easier to separate the waste into manageable fractions, and extract potentially recyclable materials, such as paper and plastics. Other systems wet the waste before drying it with hot air. As strange as this practice may seem, given that waste is already at 30 per cent moisture content, it is essentially targeted at the paper element, which in reality is quite dry. The wet paper disintegrates in the subsequent drying process, making it easier to separate out recyclable materials and turn the remainder into RDF/SRF.

Most of the MBT systems that produce RDF/SRF take out about 50 per cent of the weight of the incoming waste. This is mostly water. The moisture in the residual waste is mainly in the organic (green and kitchen waste) elements that are not segregated at source by householders. The RDF/SRF is thus half of what comes in, and since the missing half is of low caloric value (CV), the CV of the RDF/SRF is higher. The CV is still low when compared to fossil fuels (less than half that of coal), and therefore the technology that will dispose of it is basically using 'energy from waste' in the usual sense. Depending on the degree of refinement of the dried waste, and the removal of inert material, the RDF/SRF can sometimes be classified as SRF, which is governed by a European standard designed to encourage a standardisation of specification and greater use of the material as a commercially traded fuel.

Energy Recovery

How this energy is recovered and used is variable. Use of a steam-raising plant is possible unless it has been converted to gas or oil firing, in which case the mechanical plant and grate systems to handle and burn coal have often been removed long before. Large coal-fired plants burn pulverised coal, and many of the larger power companies have experimented with co-firing. However, since the physical forms of the two fuels are quite different, co-firing leads to difficulty in combustion and handling materials. Some companies have considered a dedicated RDF/SRF furnace, which feeds hot gases into the main coal-fired boiler. While this separates out the fuel-handling and combustion systems, the flue gases from the RDF/SRF are co-mingled with the main flue gases. Under current regulatory standards, this means that the whole power plant has to comply with the emissions standards prescribed in the Waste Incineration Directive (WID).

While there are slightly lower standards for co-fired situations (as compared to the standards for dedicated EfW plants – see the later section), the standards are still more stringent than those for solely coal-fired plants. Power companies are reluctant to incur the additional capital and operating costs for what amounts to just a small percentage of displacement (by thermal input) of coal. Unless the rules are changed, it is difficult to see RDF/SRF being used in large, coal-fired power plants. Another challenge!

The cement and steel industries also represent possible takers of RDF/SRF. The cement industry already takes RDF, provided it is to an agreed standard. This often involves processing the RDF to reduce moisture further. However, studies show that there are limits to the amount of RDF/SRF the cement industry can take; individual plants cannot be solely fired by RDF/SRF [2]. The steel industry also has natural and imposed limits.

As RDF/SRF has a higher CV and is also a more homogeneous material (although still quite heterogeneous in many respects), it also finds favour with certain types of dedicated EfW systems such as gasifiers. However, while it may be true to say that most gasifiers need RDF/SRF, it is not true to say that RDF/SRF needs gasifiers.

Ultimately, any RDF/SRF-burning plant is required to comply with the same air-emissions standards as any other waste plants taking a less refined fuel. Therefore, the technology employed does not need to be much different either. Apart from the use of water cooling, to cater for the slightly higher CV, a moving grate of the type seen on most municipal EfW plants is quite capable of taking RDF/SRF.

What is also true of the RDF/SRF type of MBT plant is that it is also a waste of energy. Apart from the benefit of easier extraction of certain recyclable elements from the dried waste, MBT plants absorb energy too. A typical 200,000 tonne per annum MBT plant producing 100,000 tonnes per annum of RDF/SRF uses 50–60 kWh electricity per tonne of waste. This has to be compared to a regular EfW plant that produces 500–600 kWh per tonne. Of course, the residual energy content of the RDF/SRF has to be taken into account. However, to harness that energy still requires a separate RDF/SRF EfW plant. Although such a plant is perhaps half the size of the alternative, it is more than half the cost to build and run, and is likely to be no more efficient, even if using an advanced gasification cycle.

Energy from waste (EfW)

EfW typically refers to plants, which are *dedicated* to the burning of waste materials. This includes hazardous wastes (oils, chemicals, clinical waste, etc.). Most EfW plants in the UK are built to dispose of household waste (or municipal solid waste) and sometimes commercial waste. Historically, there have been many municipal 'incinerators' across the country. However, almost all of that generation of plant (with or without energy recovery) have been closed due to ever tightening emission standards, which have made them simply unviable.

Most EfW plants use moving grates to burn the waste. These are mostly inclined beds of moving arms, which agitate the waste as it is burnt. One or two modern plants use fluidised beds. The latest example of such a system is the Allington Quarry facility recently commissioned by the Waste Recycling Group in Kent, UK. Here the waste is fed into a bed of sand upon which it is burnt.

Therein lies an important point. In most EfW plants, the waste is *burnt*; that is, it is combusted with excess air and considerable agitation to ensure complete burning (and therefore energy recovery) of the waste. This energy comes from all elements of the waste, including the large plastic component seen in modern waste streams.

Gasification and pyrolysis

Some commentators speak of gasification and pyrolysis as if they are substantially different from each other, but they are not. Gasification and pyrolysis processes also recover energy from the waste, but with less and less oxygen than combustion. Gasification heats the waste in a reduced oxygen atmosphere (about half that required to obtain combustion). This produces a complex cocktail of gases (often called a 'syngas'), which once cleaned and cooled can be used as a fuel or combusted with more air in chamber situated close behind the gasifier unit. Pyrolysis requires the waste to be heated in an (almost) oxygen-free atmosphere to produce a syngas too, which is then also cleaned before use.

In both gasification and pyrolysis, the heat required to liberate the syngas is generated by burning some of the incoming waste or the outgoing syngas it generates. There is still enough energy left to generate power. Sometimes this can be in direct combustion in a reciprocating gas engine or gas turbine, but all too often it is burnt in a combustion chamber and the heat used to generate steam for a conventional steam cycle.

Some pyrolysis systems leave a char that retains a high energy content, and this is often recovered in a separate gasification section. Indeed, some gasification plants often include an element of combustion in order to generate the required heat. Some combustion plants also have a degree of gasification in the deeper parts of the waste bed on the grate. Some syngases are claimed to be suitable for direct injection into the gas grid, but this has not been seen on a commercial scale.

Subject to the Waste Incineration Directive (WID)

Regardless of whether EfW plants are based on combustion, gasification or pyrolysis, they are all united in being subject to the WID. This imposes, among other things, stringent standards of air emissions, requiring extensive Air Pollution Control (APC) systems on the back of the heat-recovery plant. Other than that, these processes all have the hallmarks of engineering standards, and operations and maintenance disciplines, seen in 'regular' power stations. EfW is where the power industry is likely to feel most at home when it looks at operating in the waste-management space.

Apart from the fact that the fuel is difficult (not homogeneous and full of pollutants that eat steel), and power efficiency is typically two-thirds of that seen in coal-fired plants, EfW plants are much the same as any coal-, gas- or oil-fired power station. Indeed, many of the operations and maintenance staff in today's EfW plants began their careers in regular power stations.

As mentioned previously, there are systems in place to encourage waste-disposal authorities to divert waste from landfill to alternative plants. The cost of not doing this is sufficient to show that EfW plants of the combustion type are economically competitive. However, gasification and pyrolysis systems are, in the main, still at a stage of development where the capital and operating costs are higher than for combustion EfW plants. Therefore, the ROC system has been extended to include gasification and pyrolysis systems (only), so as to subsidise them through enhanced power revenues. In reality, though, gasification and pyrolysis systems are no more or less 'renewable' in the widest sense than combustion systems.

ROCs are only available for the proportion of the power generated by the biomass content of the incoming waste. Therein lies another challenge. The measurement of the biomass content of waste is not easy, and to do it with the frequency and accuracy required under the electricity regulator's current rules is difficult. As no gasification or pyrolysis plant has yet been commissioned in the UK, this situation has yet to be tested.

ROCs are also available for the biomass proportion of the power output from a combustion EfW plant if it is operating in combined heat and power mode. Again, no such plant has been built so far.

EfW opportunities

EfW (using combustion) represents the biggest opportunity in the UK for the power sector. Several of the main waste companies (including the Waste Recycling Group) are intent on taking advantage of the current policy bias to increase the amount of EfW capacity. DEFRA anticipates that up to 27 per cent of the UK's waste could be dealt with by EfW [3]. The Renewable Energy Association believes that up to 17 per cent of the UK power demand could be met by EfW [1]. The author's own view is that in order to achieve the landfill diversion targets, at least 50 new EfW plants must be built by 2020. This is on top of the 20 or so plants operating today, most of which have been built in the past 15 years. Together, they dispose of about 4.4 million tonnes per year and generate about 314 MW electricity. Only two of these plants provide heat for district heating.

Unlike regular power plants, the capital and operating costs of EfW plant are very high. For a typical plant of 250,000 tonnes per annum, the capital cost is £400–500 per tonne of capacity. In power industry 'speak', that amounts to an incredible £5.5 million per MW installed capacity, which is close to the cost of a nuclear power plant! That is not perhaps a fair analogy; a regular power plant does not also dispose of waste.

A further problem for all such plants is that in the short term the demand tends to outstrip supply. There are a limited number of suppliers capable of building plant on a lump-sum, turnkey basis. Thus, capital costs are rising, assisted in part by the effect of the construction for the 2012 Olympics in the UK.

Box 2. EfW case studies in the UK

Allington

The Allington Integrated Waste Management Facility is a 500,000 tonnes per annum EfW plant with a 65,000 tonnes per annum materials recovery facility alongside. The development process was started in 1995, and planning permission was finally granted 5 years later. The Waste Recycling Group achieved financial close on the project in 2004, and the plant was taken over from the contractor in early 2007.

Allington is a three-stream, fluidised, bed system and exports 35 MW into the local grid. It received traditional, non-recourse project finance.

Eastcroft

Eastcroft was commissioned in 1973 with a moving grate system. Originally, the APC system was basic, but was bought up to WID standards in a major upgrade in 1995. In 2005, the Waste Recycling Group undertook further upgrade work to meet the new WID standards.

Eastcroft is a two-stream plant of 100,000 tonnes per annum capacity. It supplies steam to the city council who operate a district heating system and generate up to 10 MW electricity. Work commenced in 2004 on a third stream to be built within the existing building lines. Consent of this was refused in 2006, and an appeal is presently pending, with a decision expected in early 2008.

Hull and East Riding

The Waste Recycling Group obtained planning consent in early 2007 for a 240,000 tonnes per annum EfW plant at Salt End, on the outskirts of Hull, to support an existing contract with the Hull City and East Riding of Yorkshire councils. Development work started in 2005.

The project will also be funded by non-recourse project finance, and it is hoped the facility will be commissioned in 2010/11, generating 18 MW and supplying steam to local industrial companies.

Challenges

Other than the challenges above (which are not inconsiderable), by far the biggest for EfW in the UK is planning, closely followed by planning, and coming up in third place – planning!

EfW plants built in the UK since 1990 have almost all experienced great difficulty in securing planning consent. One plant, in Lincolnshire, perhaps had an easy ride and obtained consent in the relatively short time of 6 months. The Waste Recycling Group recently succeeded in obtaining consent for a 240,000 tonnes per year plant in Hull in a similar period of time, but this was not without considerable expense. At the same time, the company was refused consent for a smaller plant in another part of the country, despite its obvious heating benefits to the local community; this highlights the vagaries and inconsistencies of the planning process. As this plant was to be an extension of the company's Nottingham facility, which is tied into the district heating system, it is hard to see why consent was not granted, and at the time of writing an appeal is under way.

The main problem with the UK planning system is that decisions are made by local politicians. These are often hard-working people who give up a lot of personal time in service to their constituents. The trouble is they are influenced by too many short-term factors, most notably being re-elected. Thus, they do not like to make unpopular decisions; and it is very safe to say that EfW planning applications are unpopular.

The unpopularity of EfW is founded on fear; fear of the unknown, driven by ignorance of the facts, and well-organised environmentalists who believe in the holy grail of ever increasing recycling levels, thus stoking up the fire (pun intended) against EfW. A large part of the opposition to EfW is pure 'NIMBY' (Not In My Back Yard) or 'NIMTO' (Not In My Term of Office) factors. For some, this is naked selfishness; for others, it is dressed up a concern for the environment, which is often short-lived or not demonstrated more widely.

In fact, modern EfW plants do not pollute. Nor do they kill local wildlife. They are not noisy, they do not ruin local buildings, and they do not smell. All of these facts can be demonstrated through the established process of an environmental impact assessment, resulting in an Environmental Statement accompanying the planning application. If they cannot demonstrate these facts, EfW applicants should, quite rightly, not be granted consent.

Refusal of planning applications often results in an appeal and a public inquiry. Here the arguments for and against are argued out courtroom style before a planning inspector. Expert witnesses are bought in to be cross-examined by barristers for applicants and opponents. The process takes a long time to set up, and a long time to conduct. It also costs all sides an awful lot of money. Often, it is the lawyers who come out on top.

If the UK government is to meet its targets for diversion of biodegradable waste away from landfill, it has to encourage a better success rate for EfW planning applications. In Ireland, such contentious decisions are taken not by politicians but by professional officers of the relevant council. Thus, the decisions are taken in a relatively political agenda-free zone. The UK would benefit from this. Recently, there have been suggestions that such a system should be introduced into UK planning system, but this is likely to be some years away [4].

Bankable projects

Other than changing the planning system, most other challenges to successful projects present relatively lower hurdles to jump. The standards to which EfW plants must comply are well understood and achievable with technologies that are proven, predictable and affordable. They are therefore bankable, and as such the finance is available for the huge capital sums involved in each EfW project. Certainly, the other problems that exist pale into insignificance when compared to the planning problem, and none are beyond being solved though conventional human perseverance and cunning.

Most EfW projects are developed and built in the same way. Mainly as a result of the planning system (but sometimes also delays in the procurement process by local authorities), most EfW projects have a long gestation period. Many are financed by limited recourse debt. This entails a particular contractual structure with which the power sector will be familiar. It requires a long-term view to be taken by project sponsors and investors. Rates of returns on investment are at utility levels, but the risks to the investment are proportionate to those returns. Payback periods are necessarily long.

Good stakeholder relationships are also key. Most notable in this respect is the relationship with the wider public, who will almost certainly be hostile to EfW proposals for the reasons outlined above. However, with careful thought and consideration, a good relationship can be established, and then the fear and hostility demonstrated in the early stages can, and often do, dissipate.

Conclusion

So all is not bad in the EfW sector. For power companies that wish to diversify from their core business, the waste sector in the form of EfW is probably a good place to be right now. The right market conditions exist, and so long as national policy continues in the current vein, there will be more investment opportunities. The water is warm!

Box 3. Top 10 tips for any large waste-management facility

- Start the planning process early.

- Check against current planning policies at regional and local level that your development fits.

- Retain good consultants with incentivised fee structures.

- Only when you are sure of your ground engage local planning officers and members.

- Engage with other stakeholders, including the public, and provide information freely and as early as possible.

- Establish websites and other communications methods at the outset.

- Be sure of your facts and figures, and be able to communicate them effectively and simply.

- Be sure to cover all aspects of the environment in preparing applications.

- Have sufficient funds (at least £300,000).

- Be prepared for a long haul.

Acknowledgements

The author is indebted to a number of colleagues in the Waste Recycling Group in the preparation and review of this chapter.

References

[1]   'Quantification of the Potential Energy from Residuals (EfR) in the UK'. Oakdene Hollins for the Renewable Energy Association and Institution of Civil Engineers, 2005.
[2]   'RDF Opportunities: Coal and Cement Industries'. Fichtner Consulting Engineers for the Resource Recovery Forum, 2004.
[3]   'Review of England's Waste Strategy. A Consultation Document'. DEFRA, 2006.
[4]   Barker, K. 'Barker Review of Land Use Planning Final Report – Recommendations'. Kate Barker for HM Treasury, 2006.

7

Financing low-level radioactive waste management: ethics, compensation and long-term governance

Erik Laes

SCK•CEN (Belgian Nuclear Research Centre),

PISA Programme (Programme on the Integration of Social Aspects into Nuclear Research)

This chapter discusses a particular dimension of 'WASTEnomics' – that of turning the long-term liabilities of low-level radioactive waste (LLW) into an ethical framework that involves 'appropriate' forms of financing and compensation for host communities of such long-term waste.

Introduction

Management of LLW seems to be a typical example of an intractable technological problem facing modern democratic industrial societies. What makes this problem so difficult to solve is not so much the technical knowledge required to build and operate LLW management facilities as the complex ethical questions it raises – especially in the repository siting phase. Different actors in the debate tend to adhere to different conceptions of (environmental) fairness and justice. For instance, the general opinion among scientific experts (geologists, engineers, modellers, assessors, etc.) in the field is that if we strive towards the best technical solution (from the point of view of objective safety and health-related criteria) and try to find the perfect site matching these criteria, people will be rationally convinced and will accept the solution proposed to them.

Setting aside the fallacy of this reasoning from a purely strategic point of view (as we will show when we discuss the Belgian experience with siting a LLW repository), this 'expert logic' tends to be at odds with other legitimate perspectives. On the other hand, local candidate communities for hosting a LLW repository will refer to a principle of autonomy and correspondingly expect to have a say in the final decision. Thus, according to the latter position, 'justice' means that local actors should have the opportunity to learn about the advantages and disadvantages of various LLW management options and, having considered these, decide on acceptance or rejection of these options.

Furthermore, there is the difficult question of relating the justification for building and operating an LLW management facility to the justification of the activities generating LLW (nuclear industry, hospitals, universities, research centres, etc.) in the first place. And – to make matters even worse – questions of intragenerational ethics (distribution of costs and benefits over current stakeholders) are compounded by questions of an intergenerational nature (distribution of costs and benefits over present and future generations). Still, these different conceptions somehow have to be reconciled in one integrated, long-term solution (it typically takes about 300 years before the radioactivity of LLW drops to background levels).

Financial aspects – whether they are related to the short-term concern of funding construction of the repository or the long-term concern of finding adequate compensation for present and future generations in the host community – are part of this 'struggle' of ethical logic, whether this is explicitly recognised or not. They have to be taken into account as one important criterion in a wider discourse on (environmental) justice framing the acceptance of a proposed LLW management solution. Conceptually, this wider discourse has to address the terms of a distributive formula expressed as follows: *Who distributes what to whom by what procedures and with what outcomes?*

With regard to aspects of financing, this formula boils down to questions regarding who has the responsibility for determining what counts as positive and negative impacts of an LLW repository and by which procedures (e.g. information gathering); what counts as an 'appropriate' form of financing (e.g. compensation for the negative impacts); and who has the authority to close the debates. The purpose of this chapter is briefly to present some key insights related to these central ethical concerns. To do so, we will begin with an exploration of these issues on the basis of some insights drawn from theories in the field of socio-psychology and ethics. Next, we will discuss a practical example of how ethical concerns were addressed in an actual siting experience in Belgium. We will conclude with a preliminary set of recommendations.

An ethical framework for long-term financing in LLW management

In this section, we place questions regarding the financing of LLW management into a broader ethical framework. Our argument unfolds as follows. Firstly, we present a 'classical' utilitarian ethical framework which supports a resolution of the debates on LLW management in financial terms. This utilitarian framework will next be used as a 'baseline' against which we will raise other ethical questions – that is, key elements of the utilitarian perspective will be subjected to a critical questioning from other possible perspectives. The issues raised include the general siting approach used, the perception of the impacts of the LLW repository on the local population, the type and form of compensation offered, links with the broader debate on (nuclear) energy policy, and the relationships between the partners involved in discussing the LLW management proposal. All in all, our argument in this section is that the siting and financing of an LLW management facility, if accepted at all by a local community, must be embedded in a jointly prepared broader 'development package' so that all residents of the local community genuinely feel 'better off'.

Utilitarianism

A 'classical' utilitarian view held by many promoters of potentially risky facilities is that the ethically correct action in a given context is the one that produces the best state of affairs – that is, that maximises 'good outcomes'. Thus, in order to establish what ought to be done in a given context, it has to be possible to compare different options (i.e. some kind of common measure has to exist), and to see which option leads to the best overall balance of negative and positive impacts. Different versions of this 'common measure' have been proposed, such as pleasure (Bentham) or the desire for happiness (Mill). More contemporary versions of utilitarianism are supported by neo-classical economic theory, which frames interactions in terms of human preferences.

After all, we all make implicit or explicit trade-offs in our daily lives, which, when made explicit, could be used to derive an estimate of how much we value our environment, our health or even our lives. Economists have developed a number of tools (e.g. contingent valuation, willingness-to-pay methods) for condensing what people value into a common monetary metric. Summing up, according to the utilitarian view, the best (rational) approach to LLW management would be the one that maximises overall well-being for society by balancing the benefits derived from activities producing the LLW (e.g. medical treatment, electricity production) against the drawbacks (e.g. health and environmental risks) on a common unit of comparison (e.g. money).

What should be compensated?

At first sight, the utilitarian framework seems reasonable enough. It is clear that the costs of building and managing the repository in agreement with health and safety regulations will have to be borne by those who benefit from the activities that generate the LLW. Furthermore, in democratic societies, it is only normal that actors engaged in a political problem-solving process both try to advance legitimate interests that should be taken into account in the problem solution, and try to question the interests and values of other actors. As these interests and value positions (at least beyond the legitimate demand for safety, health and environmental protection expressed in legal standards) are not always very transparent, having an objective calculation (e.g. a cost-benefit analysis) at hand could help to support the negotiations for a compromise solution.

But here, a problem arises. This way of arriving at a 'reasonable' solution only works when interests 'exist' as an objective reality for an economist to discover – that is, these interest have to be already formed and have to be sufficiently stabilised. To be sure, the overall welfare costs for society of different LLW management options can be established reasonably well. A cost comparison for the different alternatives (typically an extended period of temporary storage, surface disposal or deep geological disposal) will indicate the costs to be borne by the population in their role of citizens (i.e. as taxpayers) or consumers (i.e. costs included in the electricity bill or bill for medical services).

But the picture quickly gets more complicated when compensation for a possible host community accepting the burdens of a repository is considered. A host community will typically include a large variety of people with disparate interests, values and risk perceptions. For example, pregnant women, parents with children, teenagers, and highly educated men typically have very different attitudes to risk. In effect, there is a range of considerations that might be relevant: harm to human health (both physical and psychological), to the environment, to the community (e.g. loss of income from tourism), to the aesthetic value of a landscape, to the reputation of the community (e.g. social stigma), etc.

It is even very likely that these attitudes with regard to an unfamiliar technological project such as an LLW repository will be very diffuse to begin with and hence will be heavily influenced by the way this project is presented to local people. Approaching potential host communities with the promise of financial compensation can even have a perverse effect: normally, in order to increase compensation, it is necessary to depict the project as much as possible as obstructing, dangerous and damaging. This will in turn increase the resistance to the project within the community and society at large.

To conclude, utilitarian calculations seem to be useful only to have an idea of the costs to society at large, but are likely to be counter-productive when used as part of an approach to establishing compensation for potential host communities. Cost-benefit analysis can therefore be used only as a possible check on the proportionality of any form of compensation (e.g. as established on the basis of negotiations with the population suffering the possible impacts of the repository). This 'check' could warn against possible unfairness to the rest of society, as, for instance, if huge amounts of compensation are involved for only relatively minor negative impacts.

How to form a community

The argument in the preceding section can be summarised as follows: attempts to establish the right amount of compensation starting from interest positions are not likely to lead to ethically robust solutions, either because these interests are too diverse (even to the point of becoming incommensurable), diffuse or affected by a negative spiral of raising the bid. We are not suggesting here that host communities should not be compensated for taking their part of the burdens of an LLW facility. But we are suggesting that an *initial* framing of interactions with a potential host community in terms of market-type negotiations tends to be counter-productive. Furthermore, establishing relationships with a potential host community on the presupposition that this community consists of rationally calculating egoists raises serious ethical concerns, for it elides the central ethical question concerning the kind of community we can form including this new and strange (in the eyes of the local population) technological project.

In fact, there is a danger of talking about a potential 'host community' (as we have done up until now) in a way that implies that this is a homogeneous group. In contrast, the 'host community' is not simply a given but has to be formed in the process of looking for solutions to the LLW management problem. Difficult questions have to be answered. For instance, where does the 'host community' end? Is it delimited by an administrative boundary (e.g. a municipality) or by a cultural identity (e.g. a region)? And if the potential repository site is near the administrative border of a municipality, does the 'host community' include the neighbouring municipality?

The point here is not to answer these questions but simply to draw attention to them. In any case, it can be argued that it will be inappropriate to consider the question of fair compensation unless the 'host community' has been, and can be seen to have been, chosen by a due process (genuinely open; guided by criteria that are clear, comprehensible and reasonable; informed by dialogue and debate with relevant stakeholders, etc.). This is very important for possible compensations not to be perceived as bribery – that is, there has to be a good reason, *independent of the compensation offer*, for selecting that particular site as a host. Furthermore, socio-psychological research has shown that it is easier to form a community of dialogue partners if people understand that the proposed facility meets a genuine need (i.e. that it is needed from a 'societal perspective'), and if they understand that the consequences of doing nothing about the problem would be worse for everyone. Of course, a key question here is to what extent the 'civic' or 'ethical' motivation actually exists.

A further question that arises is whether and under what circumstances it would be ethical to approach a potential host community on the basis of the existence of such civic motivation. As will become evident from our discussion of the Belgian case, this civic motivation of resolving a common problem is more likely to be present in nuclear communities (i.e. communities already hosting a nuclear facility on their territory), so from a tactical point of view, successful negotiations are more likely in such communities. But does it follow, then, that such communities *ought* to take on the responsibility of hosting an LLW facility from an ethical point of view?

In any case, here again we can point out the possible perverse effect of approaching a potential host community with an offer of financial compensation. There is evidence that, where a community has an intrinsic reason – such as an expressed willingness to contribute to the 'common good' – to accept a potentially hazardous technological activity on their territory, offers of financial compensation reduce rather than enhance this willingness. A possible explanation is that the financial offer transforms an ethical relationship into a monetary one in a way that undermines the initial motivation.

What is appropriate compensation?

If a potential host community has been chosen by due process, and conclusions about what counts as positive and negative impacts of the LLW management facility have been reached by a dialogue process with local stakeholders, judgements will next have to be made about what form compensation should take (given the nature of the identified impacts). In effect, the form of possible compensation matters greatly for the kind of relationships that will be established between the actors involved in negotiations, including the ethical norms governing them.

Following socio-psychological theory, three fundamental and distinct forms of social relations can be distinguished: communal sharing, equality matching and market pricing. Under communal-sharing relations, goods are collectively owned (and people are prepared to contribute a great deal without expecting much in return). Under equality-matching relations, a contribution or favour would be expected to be matched in kind. Finally, under market-pricing relations, a contribution would typically be expected to be paid for. For our purposes, the important point to retain is that in the case of siting an LLW management facility, someone who frames the siting problem as a communal-sharing problem (i.e. acting for the 'common good') and is offered money as compensation (i.e. a market-pricing mechanism) will react with indignation.

The theory suggests that in the process of trying to identify 'appropriate' compensation, attention should be paid to the way in which members of a potential host community understand the relationship between themselves, the LLW management facility, and the people offering compensation. And in the case of LLW management, the nature of these relationships (e.g. relationships between those agreeing to host the facility and their descendants) is more likely to be of the 'equality-matching' or 'communal-sharing' type (as we will see in the next section, an investigation of the Belgian case lends credence to this idea).

Issues of wider concern

A final set of ethical questions relates to issues of wider concern. Although siting an LLW management facility is a local activity, it is clear that this process has national repercussions. For instance, proponents of nuclear power will probably construe success in siting an LLW management facility as a step in the direction of greater public endorsement of the nuclear option. Conversely, adversaries (e.g. environmentalists) will generally try to thwart the efforts to site an LLW facility in order to demonstrate that the nuclear industry cannot solve one of its most difficult problems.

Thus, the question of where and how an LLW facility is to be hosted will be impossible to separate from the question of whether radioactive waste should continue to be produced in the first place. The siting dynamic, including its financial aspects, will feed into the dynamic of the wider debate on nuclear power. For instance, perceptions of the fairness of the decision to produce (or continue producing) nuclear power will have an impact on the local siting debate. Thus, as a general rule, it is important that the local siting debate is in some way connected to the broader context of electricity generation and demand.

Case study: the Belgian 'partnership approach'

In this case study, we investigate how some of the above-mentioned ethical choices regarding financing and compensation have been negotiated in an actual policy setting.

Background

Since 1980, radioactive waste management in Belgium has been entrusted to the Agency for Radioactive Waste and Enriched Fissile Materials (NIRAS). Following the international moratorium on the dumping of LLW in the north Atlantic Ocean, NIRAS took charge of the conditioning and storage of this type of waste (entrusted to its industrial subsidiary, Belgoprocess). Over the years, NIRAS investigated three disposal options for LLW: coal mines or quarries, shallow land burial and deep geological disposal.

In 1990, NIRAS concluded that the surface disposal option was the most favourable from the point of view of technical feasibility, safety and cost. This option was subsequently studied and developed in greater detail. In the meantime, the debate on the 'back-end' of the fuel cycle had become more intense, as the result of (international) political developments. Because sea dumping was permanently banned in 1993 (and export of radioactive waste is not permitted in the European Union), an inland solution for the disposal of LLW had to be found, leading to severe political friction and NIMBY-type reactions to the top-down approach first employed in 1994 by NIRAS (publication of a report that acknowledged 98 potential zones in Belgium for disposing at least 60 per cent of the LLW on the basis of technical criteria only). This in turn led to a new strategy of allowing participation of local stakeholders and citizens in so-called local partnerships with NIRAS, based on the concept that every party that could be directly affected by the decision concerning LLW management should have an opportunity to express its opinions.

As a result, NIRAS had to limit its search for potential hosts to the existing nuclear zones in Belgium (i.e. Doel, Fleurus-Farciennes, Mol-Dessel and Tihange) and to local towns or villages that volunteered in preliminary field studies. Four communities in nuclear zones (Mol, Dessel, Fleurus and Farciennes) responded positively, and three partnerships were erected (STOLA in Dessel, MONA in Mol, and PaLoFF in Fleurus-Farciennes) between 1999 and 2002. The partnerships have a general assembly, comprising representatives of all interested organisations in the community in question. The general assembly represents all participating organisations, decides on the main objectives and boundaries of the discussions, and appoints an executive committee. This executive committee is in charge of the day-to-day management of the organisation, budget supervision, coordination of working group activities, etc.

MONA, STOLA and PaLoFF also each have two project coordinators and four working groups. These working groups all debate a specific topic. In the case of MONA, these include implantation and design (I&D), environment and health (E&H), safety (S) and local development (LD). The I&D working group is the most 'technical' one, as it discusses the repository design and the site location. The E&H working group looks at the possible impacts on human beings and the environment. The S working group discusses safety and emergency issues. Finally, the LD working group debates the possible 'added value' of the proposed repository to the community. After some years of debate, all partnerships have in the course of 2005 published their conclusions and recommendations in final reports.

All have agreed to the construction of an LLW repository on their territory, provided that certain conditions are met (e.g. improved emergency planning, maintenance of nuclear expertise in the region, monitoring of levels of radioactivity, etc.). In the case of MONA and STOLA, these final reports have been endorsed by the respective municipal councils; in the case of PaLoFF, it was rejected – hence, Fleurus-Farciennes has withdrawn from the site-selection procedure. It was up to the Belgian federal government to choose between the two remaining candidate communities, and it chose Dessel (June 2006). The government also stated in its decision that neighbouring communities must be able to participate in the following phase of the repository project. Now that the site has been selected, the role and coordination of the activities of both partnerships will have to be reviewed.

Discussion

Within the limits of this chapter, it is of course impossible to discuss in detail the pros and cons of the Belgian 'partnership approach'. Let us just quickly mention some of the most conspicuous characteristics that also help to develop an understanding of issues of financing and compensation. The following characteristics seem to stand out:

- The partnerships were initiated by NIRAS – the organisation responsible for finding and implementing radioactive waste-management solutions, and thus with an interest in finding a quick solution (taking into account, of course, all necessary safety and environmental standards) to the LLW management problem after some costly previous failures.

- The nuclear communities that applied for a partnership with NIRAS did not figure in the original list of 98 'ideal' zones for a surface LLW repository. This change of policy was justified only in rather vague terms – thus, in the final report published by MONA (p. 21), we read that NIRAS was no longer looking for the 'ideal' site for the 'ideal' concept, but rather for a 'suitable' site for a 'befitting' concept. This ambiguous phrasing in any case implies a move away from strictly technical criteria in site selection to 'other' criteria without giving further specifications, thus raising confusion about the motives behind this change (but see the 'Ethical framework' section for some of the motives that might have influenced the choice for nuclear communities).

- The partnership approach puts the emphasis on local (i.e. municipal) capacities to integrate the siting project in a wider development project (thus, priority is given to local actors in participation). Doing so limits the empowerment of citizens to the negotiation of a local order – that is, a contextually appropriate solution.

- Furthermore, the partnership approach is clearly based on a clear-cut and traditional boundary between the 'collective interest' represented by public authorities and 'local interests' represented by citizens. In effect, federal authorities (e.g. the regulator) and/or government representatives were only involved in the process 'at a distance'.

- The partnerships open the discussion on expert knowledge and techniques;

this enables a move away from a 'rhetoric of fear' based on risk perception and media amplification. On the downside, however, the partnership approach runs the risk of turning the participants into an 'expert club' themselves, and thus losing the link with the vernacular understanding of risk in the community.

Concerning financing and compensation, we will discuss the case of MONA (as we are most familiar with the workings of this particular partnership), and in particular the activities of the LD working group. To begin with, a distinction must be made between the short-term funding of the partnership activities and the long-term funding of the LLW repository. The partnership activities were funded entirely by NIRAS (which in turn is funded by the producers of radioactive waste according to the 'polluter-pays principle') – e.g. MONA received a yearly donation of about 250,000 euros, and two one-off budgets of some 75,000 euros for carrying out socio-economic studies and for developing a proposal for an integrated repository concept. The partnerships were free to use this budget according to their own judgement. Issues of long-term financing and compensation for the possible siting of an LLW repository were mainly addressed in the discussions taking place in the LD working group of the MONA partnership.

Initially, the activities of the LD working group were oriented towards the identification of a list of projects that represented an 'added value' for the community – such as projects concerning employment creation, health, the environment, or improved mobility. The group derived a list of selection criteria for setting priorities – projects should serve the 'common good', be committed to social goals, be ecologically sound and sustainable, etc. It is clear that the LD working group in its initial deliberations was trying to identify 'well-matched' compensations for accepting an LLW repository according to the equality-matching principle discussed above.

According to this logic, the host community is entitled to compensation to restore the balance of justice; hence, the list of priority compensatory projects reveals an intention to establish an extended moral community (e.g. by promoting employment, improving social conditions) or re-establishing links with 'nature' (e.g. by promoting ecologically sound and sustainable projects). However, it was never really made explicit what the local community should be compensated for – in fact, the word 'compensation' was never mentioned in the 'official' partnership discourse! Instead, it stated that partnerships should aim to construct an 'integrated repository concept with added value for the local community'. This finding strongly suggests that even the mere act of balancing or trading off positive against negative aspects reeks of unethical behaviour, at least in the case of LLW management.

And the issue gets even more complicated when future generations enter into the picture. In a second phase, the LD working group realised that projects should have 'added value' not only for the present generation, but also for the future ones. Thus, the LD working groups changed their opinion and considered it to be inappropriate to identify a list of projects based on current preferences (for preferences could change in the future); instead, the LD working group now took up the idea of establishing an autonomous, long-term fund for the financing of future projects (still subject to guidelines on the suitability of these projects, as laid out in the mission statement of the fund).

Furthermore, the establishment of this fund was connected to a demand for more empowerment and autonomy for the local community, for following up not only the issue of LLW management, but also other nuclear activities in the region (e.g. interim storage of high-level waste, transport of radioactive material, monitoring of experiments in an underground laboratory, etc.). 'Civic' motives (e.g. empowerment, autonomy) were now advanced to restore the balance of justice, even challenging the original mandate (i.e. discussing LLW management) and municipal scope of the partnership approach (as mentioned, the neighbouring municipalities of Mol and Dessel want to work together in the future). The bargain seems to be acceptable to NIRAS: in Dessel, the follow-up to STOLA (called STORA) is already active, while at the time of writing of this chapter there was an agreement in principle between NIRAS and MONA to form a similar follow-up committee in Mol.

But again, the details of the compromise solution have not been made public – most conspicuously, the financing of the fund is discussed only in the most general terms (it must be a 'legally binding agreement between the competent persons'). Hence, we can only conclude that the final (financial) settlement achieved is formulated in everyday language, where goals, means and outcomes are stated in rather vague and possibly ambiguous terms. But this seems to be a successful way of moving forward in a very complex context charged with references to justice and ethics.

Recommendations

So, what can be learnt from our previous discussion? It is clear that questions about 'appropriate' forms of financing and compensation for communities that agree to host an LLW management facility need to be considered in conjunction with a broader set of questions about the relationships between the host community and wider moral communities (e.g. future generations, the rest of society). We hold that the following points should be taken into consideration when pondering the difficulties of identifying ethically acceptable ways forward:

- The initial site-selection process should ideally be open, transparent and informed by a public debate (e.g. by well-informed representatives of stakeholder organisations and/or public involvement mechanisms).

- This process should include reference to the broader context of energy demand and generation, and should not be focused on nuclear energy in isolation.

- Potential host communities should be approached with an 'open problem framing', in order to establish a sense of shared problem ownership.

- Citizens should be allowed to have a voice in the negotiations on acceptable forms of financing or compensation, rather than leaving this task to political appointees (who tend to look at this from a short-term point of view).

- Ideally, those offering compensation should not have (or should be seen not to have) anything to gain from finding a solution to the LLW management problem. Therefore, we recommend that the negotiations on compromise solutions be facilitated by an organisation that is genuinely independent of the nuclear industry.

- Compensation should be proportionate to negative (and positive) impacts. While a broad cost-benefit analysis can contribute to specifying the meaning of 'proportionate' in this context, we do not recommend framing the initial debate setting in terms of a compensation 'threshold'.

- Instead, the type of compensation should reflect the type of relationships that the local community wishes to establish with the LLW repository and those responsible for building and managing it.

- One-off forms of compensation should be avoided, as these tend to neglect the interests of future generations (who have not benefited from the activities generating the radioactive waste). Instead, a legally anchored fund that protects the interests of future generations should be established.

We contend that solutions to this intractable problem can be found only by treating in a single process the technical LLW management project, its placement in a given environment, and its possible integration in a socio-economic fabric (including financing).

.

Section 3:
Waste Management
and Waste-to-Energy

8

Towards environmentally, financially and socially sound waste-management practices in Asia

A. Prem Ananth and C. Visvanathan
Environmental Engineering and Management Program, Asian Institute of Technology, Bangkok, Thailand

This chapter discusses a key dimension of 'WASTEnomics' – that of ensuring environmentally, financially and socially sound waste-management practices.

Introduction

Though industrial and economic development in Asian countries has been progressing on a par with some developed countries, management of solid waste is still an issue warranting immediate attention. National governments and local administration units face constraints in terms of collection and disposal, and they require support in many aspects, including technologies, policies, institutional frameworks, investments and human behavioural changes.

Not all is bad in waste management. Scientists, technocrats, policymakers, businesses and financial institutions have been trying to address the issue in all possible ways. However, most of the initiatives have aimed at the disposal phase of the problem rather than tackling it at the source (as in zero waste practices), especially in most of the developing Asian countries. This chapter outlines the current trends of solid waste management in some Asian countries and broadly discusses some initiatives aimed at tackling the problem at source.

The chapter first introduces the reader to the Asian waste situation, especially in waste generation, composition, collection and associated financial aspects. Subsequently, a short discussion is presented on partnerships in the waste sector and an overview of the prevailing legislation on waste management in some Asian countries. The remaining sections of the chapter concentrate fully on the three Rs of waste management: *reduce*, *reuse* and *recycle*, a pragmatic approach to tackle the waste crisis at the source.

Overview of the current situation

Most cities in the Asian and the Pacific region are struggling to cope with overflowing landfills and rapidly increasing volumes of municipal waste resulting from increased economic productivity and consumption. The consequence is that municipal governments spend up to 50 per cent of their budget on solid waste management operations, much of it on collection and transfer [1].

Government spending on environmental protection amounts to less than 1 per cent of gross domestic product (GDP), while the World Bank calculates that neglect of the environment is costing an average of 5 per cent of GDP in Asian countries. For example, China is believed to be losing as much as 10 per cent of its national income to pollution, while for India it is 5–6 per cent.

Japan, with an average annual outlay of 1.8–2 per cent of GDP, spends the most on environmental protection, followed by South Korea with 1.3–1.6 per cent, Singapore with 1.2–1.5 per cent, and Taiwan with 1–1.2 per cent of GDP. Among the developing countries, Vietnam spends only about 0.1–0.3 per cent of GDP, and China, Indonesia and the Philippines spend 0.5–0.7 per cent. Malaysia and Thailand both invest almost 1 per cent of GDP on the environment.

The most neglected pollution sector is generally assumed to be the land environment, which sustains the biggest impact from inadequate waste disposal systems and is often more difficult to monitor [2]. Consumption and production statistics from the rapidly industrialising Asian countries, especially China and India, clearly demonstrate the challenges faced by the government in handling the solid waste management issue.

Solid waste generation and its composition in developing Asian countries vary widely due to population and economic conditions. For example, Thimpu, Bhutan, generates about 36.7 tonnes/day while in Dhaka, Bangladesh, it is as high as 4,000 tonnes/day.

Urban India generates about 120,000 tonnes of garbage every day [3], of which Delhi contributes 5,900; Greater Mumbai, 5,300; Chennai, 3,000; Kolkata, 2,600; Hyderabad, 2,100; Bangalore, 1,600; and Ahmadabad, 1,300. In Kathmandu, Nepal, the daily garbage flow is 944 m³, approximately 300 tonnes [4], whereas in Colombo, Sri Lanka, it is about 2,900 tonnes/day [5].

Thailand generated approximately 40,000 tonnes/day in 2005, totalling 14 million tonnes in a year. Of this, close to 21 per cent was from the Bangkok metropolitan area and about 32 per cent from other municipalities and Pattaya City. The rural areas and subdistricts contributed about 47 per cent.

The rate of solid waste generation in the Philippines is comparable to other low to middle-income countries. The National Capital Region alone accounts for 23 per cent while the Southern Tagalog Regions contribute 13 per cent of the total waste generated annually. The Asian Development Bank estimates that about 6,700 tonnes of waste is generated every day in Metro Manila alone.

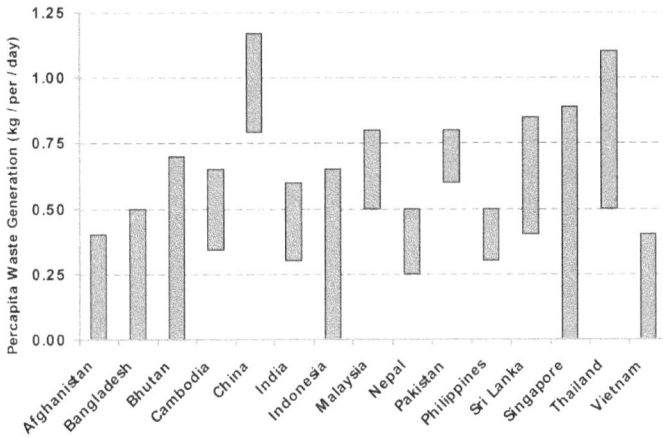

Figure 1. Waste Generation Rate in some Asian countries

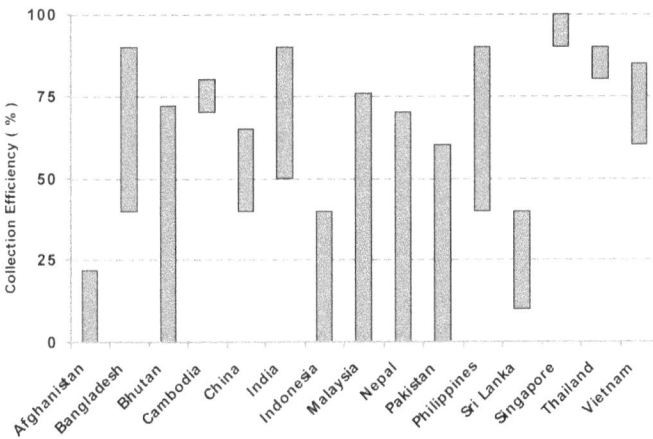

Figure 2. Waste Collection Efficiency in some Asian countries

In Vietnam, urban areas contain a mere 24 per cent of the country's population, but produce over 6 million tonnes (50 per cent) of the country's municipal waste.

Singapore has been able to maintain a near-constant waste generation rate of 2.5–2.7 million tonnes in the past decade. Not much variation is seen in the generation rate. In 2006, about 2.56 million tonnes of waste was disposed of, of which 57 per cent was generated by residential premises, food centres and markets. Commercial and industrial premises accounted for the remaining 43 per cent.

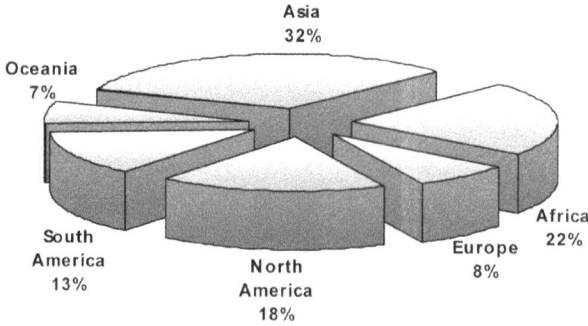

Figure 3. Land Area of Continents

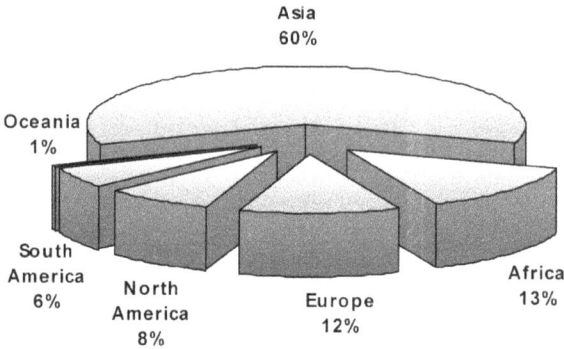

Figure 4. Population of Continents

In Malaysia, the amount of solid waste increased from 16,200 tonnes/day in 2001 to 19,100 tonnes/day in 2005. The biggest generators are the states of Selangor and the Federal Territory, both of which are highly urbanised and industrialised areas. These states account for 36 per cent of the municipal solid waste (MSW) generated in Malaysia.

Figures 1 and 2 present data on the per capita waste generation and collection efficiency in some Asian countries. Evidently, Asian countries do not generate as much waste as the other developed countries. It is the relatively higher population that intensifies the problems of waste management. Furthermore, the resources available to tackle the issue in these countries are limited.

If we look at the big picture, the key commonality among most Asian countries is that waste generation levels are constantly rising, for reasons that are many and widely debated. Rapidly increasing population levels, changes in lifestyle, affluent consumerism, urbanisation, and industrialisation are the factors causing the present upward trend in waste generation.

Asia faces more acute pressure on land than any other region in the world. Figures 3 and 4 show that, with only 32 per cent of the world's land area, the Asian continent is home to 60 per cent of the world's population. This is a clear indicator of the stress on the environment in the region, most of which is from the consumption of resources and disposal of waste. With more population and more demands to satisfy its daily needs, Asia needs more resources, eventually resulting in more waste.

Collection is the main aspect as far as management of solid wastes is concerned. Governments and municipal bodies in developing countries face difficulties in many forms. With limited budgets and manpower, municipal authorities in Asia are unable to cater for the ever increasing needs of the society. Municipal governments are usually the responsible agency for solid waste collection and disposal. However, in most developing Asian countries, the magnitude of the problem is well beyond the ability of any municipal government.

In Asian developing countries, small, poor-quality metal or plastic containers, or enclosures and waste platforms, which overflow before the scheduled collection time, are used for waste collection, rather than well-monitored bins.

Transfer and transport systems are an important factor. Most Asian developing countries have narrow streets and lanes that hamper the movement of vehicles. It is common that the larger collection vehicles do not cover such areas. Modern, specially designed, smaller collection vehicles are needed to service these areas, but they are too expensive for most municipalities. The overall maintenance of the vehicles and their availability at the scheduled time are two factors that influence collection efficiency.

In essence, the key factors of poor collection performance are inadequate resources and lack of attention by the government. Added to this is the sluggishness of the authorities in comprehending the issues that arise out of inappropriate solid waste management [5].

In addition to the various levels of government, businesses and general communities need to be more involved in waste management (Box 1).

Box 1. Snapshots of solid waste management in Asian countries

Bhutan

As mandated by the Municipal Act 1999, waste collection and disposal responsibility rests with the municipal authorities. Thimphu City Corporation is solely responsible for the MSW management of Thimphu City. The corporation spends around two million BTN (about US$55,000) annually on the collection and disposal of waste. A formal waste segregation system is yet to be initiated.

China

The total MSW generated annually in China is 120–136 million tonnes. Of the total solid waste, nearly 60 per cent is generated by 52 cities, each having a population of over 0.5 million. China has recently surpassed the USA as the world's largest MSW generator; nevertheless, it has demonstrated little waste reduction effort. With respect to the amount of MSW generation and its composition, the waste treatment and disposal methods are poor. For instance, over 70 per cent of the generated waste is disposed of in open dumps, and only about 20 per cent is being composted.

India

Except in metropolitan areas, MSW management is the responsibility of a health officer, assisted by the engineering department in transportation. The activity is quite labour intensive, and 2–3 workers are provided per 1,000 residents served. Municipal agencies spend about 5–25 per cent of their budget on solid waste management; that is, Rs.75–250 (US$1.8–6.25) per capita per year. Normally, a city of one million population spends around Rs.100 million (US$2.5 million) on this activity. Despite this huge expenditure, it is widely noted that services are not provided to the desired level [6]. The urban local bodies spend approximately Rs.500–1500 (US$12.5–37.5) per tonne on collection, transportation, treatment and disposal. About 60–70 per cent of this amount is spent on collection, 20–30 per cent on transportation, and less than 5 per cent on final disposal. Of the total waste collected, over 95 per cent is dumped and less than 5 per cent is composted.

South Korea

South Korea's solid waste from domestic and economic activities has substantially decreased since the introduction of the volume-based waste fee system in 1995. The system resulted in a savings of about US$7.7 billion as a consequence of a waste reduction of about 61 million tonnes and an increase in recyclables collection of about 28 million tonnes. Before the implementation of the volume-based waste fee system in 1994, the solid waste generation was 1.33 kg/capita per day; this decreased to 1.07 kg/capita per day in 1995, and further decreased to 1.04 kg/capita per day in 2003.

Malaysia

In Malaysia, local authorities spend up to 60 per cent of their annual budget on waste management. The government of Malaysia has privatised solid waste management and has adopted an integrated system. Since 1997, two companies, Alam Flora Sdn Bhd and Southern Waste Management Sdn Bhd, have been engaged in management contracts with the local authorities within their respective concession areas. One private waste concessionaire has undertaken a pilot project to compost green waste in its landfill. The concessionaire has undertaken to compost 8 per cent of the collected waste.

The Philippines

It is estimated that Metro Manila spends over PHP3.5 billion (US$80 million) annually on the collection and disposal of solid waste. The League of Cities of the Philippines survey in July 2005 revealed that the efficiency of garbage collection in cities averages 40 per cent in the political districts in which about 56 per cent of the cities' population is served. Quezon City reported a collection efficiency of 99 per cent in 2005.

Singapore

In 2006, Singapore generated about 2.6 million tonnes of waste, which was disposed primarily through incineration and landfill. Singapore is divided into nine geographical sectors, in which Public Waste Collectors (PWC) and General Waste Collectors (GWC), licensed by the National Environment Agency, carry out solid waste collection. Industrial and commercial premises are served by licensed GWCs. In addition to conventional waste collection services, the PWCs are also responsible for providing recycling services to residents in their sectors. Currently, about 45 per cent of the generated waste is incinerated, while 4 per cent, consisting mainly of non-incinerable waste, is sent directly to landfills. The remaining 51 per cent of the waste is recycled.

Sri Lanka

Apart from the financial support from the Provincial Council and/or the Treasury, other revenue sources of local authorities are acreage tax, licence fees, assessment tax (property tax), fines and service charges. The Municipal Council, Urban Council and Pradeshiya Sabha spent approximately 14, 20 and 12 per cent, respectively, of their budgets on waste management [7]. Most of the budget is spent on salaries to staff and labourers, fuel for vehicles, and vehicle maintenance. The expenditure on final disposal is low. In Sri Lanka, no direct fee is levied on households for waste collection or disposal. Instead, the cost of waste management is covered by property tax.

Japan

Over 51 million tonnes of municipal waste contributes to the waste stream, at an average generation rate of nearly 1.14 kg/capita per day in Japan. The waste is segregated into general waste and recyclable waste, and it is collected separately. The collected general wastes are sorted as combustible and bulky. While the combustible wastes are sent directly to waste incinerators, the bulky items are compacted before they are sent to landfills.

Considering the growing need to accelerate recycling as a means of solving the waste management crisis, the government of Japan established Eco-Towns in 1997. In addition, various laws emphasising recycling and reuse were enacted. As a result, the actual amount of waste sent to landfills has been greatly reduced since the year 2000, and Japan has been able to achieve the highest level of resource productivity.

Composition and characteristics

An overview of the composition of waste in some Asian countries is presented in Table 1. Evidently, about 50–70 per cent or more of this waste is organic, the remainder consisting of paper, plastic, glass, metal and other inert materials. It is essential to mention that the composition differs by the economic level of the country and other factors such as geographic location, energy sources, climate, living standards, cultural habits and, especially, the country's definition of MSW. However, one important aspect is that a significant variation, sufficient to prevent any generalisation of the waste categories, exists, thus preventing the recommendation and adoption of a single technology to handle them.

Table 1. Characteristics of waste in some Asian countries

Country	Organic	Paper	Plastic	Glass	Metal	Others
Bangladesh	70.0	4.30	4.70	0.4	4.6	16.0
Cambodia	65.0	3.8	13.2	4.9	1.0	12.1
China	35.8	3.7	3.8	2.0	0.4	54.3
India	45.0	7.0	4.0	2.0	2.0	40
Indonesia	70.2	10.9	8.7	1.7	1.8	6.7
Japan	17.0	40.0	20.0	10.0	6.0	7.0
Korea	31.0	27.0	6.0	6.0	7.0	23.0
Laos	54.3	3.3	7.8	8.5	3.8	22.3
Malaysia	43.2	23.7	11.2	3.2	4.2	14.5
Myanmar	80.0	4.0	2.0	0.0	0.0	14.0
Philippines	41.6	19.5	13.8	2.5	4.8	17.8
Singapore	44.4	28.3	11.8	4.1	4.8	6.6
Thailand	48.6	14.6	13.9	5.1	3.6	14.2
Vietnam	49.0	2.0	16.0	7.0	6.0	20.0

All values are expressed in percentages and on a wet-weight basis.

Financing and partnerships

Investment and budgets are the main barriers to proper solid waste management as far as developing Asian countries are concerned. Proactive policies, reforms in fiscal measures and adopting only innovative economic instruments can help to enhance the financial capacity of local municipal and urban bodies. Solid waste management authorities, essentially the local administration units, in Asia should move towards recovering their costs by levying fees for the various services provided. The ideal example proven successful in some developed Asian countries is generally either a direct fee based on waste volume or an indirect fee derived from another levy, such as property tax.

Recently, Asian cities, applying the polluter-pays principle, have started to levy charges on citizens to formalise the construction and operation of waste disposal facilities and provide incentives to residents to separate their waste into different types according to the treatment requirements.

Private sector participation, by way of partnerships with local bodies, is slowly penetrating the waste sector in many countries. For example, Chennai, the fourth largest city in India, was the first to outsource MSW management services to a private agency, CES Onyx. The scope of the contract included maintenance activities such as sweeping, collection, storing and transporting the solid waste from three zones. Before contracting, solid waste management was primarily the function of the local municipal body.

Since it started work in March 2000, CES Onyx has taken credit for modernising garbage clearance in the city. The company won a global bid floated by the corporation of Chennai, the local administration body. The private operator recruited over 2,000 workers, trained them in modern garbage clearance, provided them with safety gear, and brought about a scientific method of garbage clearance in the city. Since then, the company has been clearing close to 1,100 tonnes of mixed waste, including construction debris, every day.

The corporation has been paying charges for clearing garbage and debris from three zones and transporting it to the dump. The tipping charges increased from Rs.648/ tonne (US$16.20) in the first year of operation to Rs.1,212/tonne (US$30.30) in its seventh year. Overall satisfaction levels are reported to be higher in areas cleaned by CES Onyx than by the urban municipal body.

In East and Southeast Asia, private sector participation has increased in the form of lease and concession contracts in construction and operation of facilities for solid waste disposal. For example, in Malaysia, waste collection and the construction/operation of solid waste disposal facilities have been transferred to private companies through special contracts. Similar lease and concession contracts have also been adopted in many other countries such as the Philippines, Thailand, and Singapore. However, private sector participation is often not motivated as a result of difficulties arising in acquiring land for operation sites due to local opposition. The foremost reasons projected by the local residents are safety risks due to increased traffic and movement of vehicles, spread of vector-borne diseases from the facilities, possible groundwater contamination, and eventually loss in commercial value of property. Despite all these factors, private sector participation in solid waste collection, transport and recycling has increased.

Policies and legislations related to waste management

In many developing Asian countries, the current regulation system has enormous room for improvement; the system in practice is either imperfect or insufficient, or it does not fit the present requirements. A strategic approach involving the amendment of current laws and regulations, improving the current management systems, and introducing classified collections is greatly needed to tackle the issue systematically. Above all, effective implementation of these strategies is essential to solve the problems of waste management. Table 2 provides an overview of legislation that directly or indirectly addresses waste management issues in some Asian countries.

An understanding of the prevailing legislation in developing Asian countries clearly indicates that management of waste has been acknowledged as an issue. However, the level of action envisaged is still not satisfactory. In some countries, the prevailing legislation attempts to address the issue by following a command and control approach. Source reduction or proactive approaches have not been adequately envisaged. Though the legislation in some countries is ambitious, it fails to address issues related to investment and the funds needed to implement schemes and programmes.

The mere presence of ambitious legislation does not itself solve the problems. Effective implementation must go hand in hand. After all, generation of waste is highly linked to human behaviour and lifestyle. Although most Asian countries have applicable legislation, they still lack the institutional framework and human resources for effective implementation. In fact, adequate institutional framework is linked to investment and budget constraints.

Table 2. Waste-management-related legislation in some Asian countries

Country	Legislation
Bangladesh	Environment Conservation Act, 1995
	Environment Conservation Rules, 1997
	Environment Policy, 1993
	Draft Urban Solid Management Handling Rules of Bangladesh, 2005
	Draft Medical Waste Management and Handling Rules, 2005
Bhutan	Environmental Code of Practice for Solid Waste Management, 2000
	Environmental Code of Practice for Hazardous Waste Management, 2002
	Policy Framework for Solid Waste Management, 2006
	National Strategy and Action Plan for Integrated Solid Waste Management, 2007
Cambodia	Law on Environmental Protection and Natural Resources Management
	Sub-Decree on Solid Waste Management
China	City Appearance and Environmental Sanitary Management Ordinance, 1992
	Regulations Regarding Municipal Residential Solid Waste, 1993
	Prevention and control of Pollution by Solid Waste, 2004
	Regulations Regarding Municipal Construction Waste, 2005
India	The Hazardous Wastes (Management and Handling) Rules, 1989
	The Bio-Medical Waste (Management and Handling) Rules, 1998
	Solid Wastes (Management and Handling) Rules, 2000
	The Batteries (Management and Handling) Rules, 2001
	National Environmental Policy, 2005
Indonesia	Environmental Management Act, 1997
	Domestic Solid Waste Management Act, 2006 (draft)
Malaysia	Local Government Act, 1976 (Act 171)
	National Strategic Plan for Solid Waste Management
	Action Plan for a Clean and Beautiful Malaysia

Country	Legislation
Nepal	Solid Waste Management and Resource Mobilization Act, 1987
	National conservation Strategy, 1988
	National Waste Management Policy, 1996
Philippines	Integrated Environmental Protection and Natural Resources Management Policy 25
	Presidential Decree 1586 The Environmental Impact Statement (EIS) System
	Solid Waste Management Act (RA 9003)
	Ecological Solid Waste Management Act of 2000
	Toxic Chemicals and Hazardous Waste Management (RA 6969)
	The Philippine Agenda 21
Sri Lanka	National Strategy for Solid Waste Management
	National Integrated Waste Management Plan
Thailand	Enhancement and Conservation of National Environmental Quality Act, B.E. 2535
	Solid Waste, Night Soil and Hazardous Waste Management
	Toxic Substance Legislation
	National Strategy for Environmental Protection
Vietnam	Law on Environmental Protection, 1994
	Directive on Urgent Measures on Solid Waste Management in Urban and Industrial Areas, 1997
	Decision on Hazardous Waste Management, 1999
	National Plan for Environment and Sustainable Development, 1999

An overview of the current situation with perspectives on waste generation, collection, characteristics and composition raises different thoughts. Most of the Asian countries are in a serious phase requiring immediate action. Waste generation and management is an issue with social aspects. Policies and legislation play an important role when it comes to people-centric aspects of management. An understanding of the prevailing policies makes clear that in most, if not all, Asian countries, appropriate policies do exist. However, what the policies envisage and how they will be achieved in real life is still the million-dollar question. Given the constraints in terms of treatment technologies, investments and other factors, it is essential that strategies able to circumvent these constraints, over time, need to be developed. Is there any such strategy in place that could be pursued with minimum effort?

Reduce, reuse and recycle – the three Rs of waste management

Undoubtedly, solid waste generation in Asia has displayed an increasing trend on a par with the pace of population growth, urbanisation and industrialisation. Sufficient evidence exists to argue that managing solid waste has been simply transporting the waste to distant dumps, literally moving it out of sight. Even in this case, often only a fraction of the waste is properly collected and transported. Most often, it is burnt to reduce the volume, minimise the attraction of animals and vermin, and retrieve the recyclable materials. The disadvantage of the open dumping practice is the adverse effects on the groundwater sources. Leachate, the water that collects contaminants as it trickles through the dumped waste, extracts dissolved or suspended materials, including potentially harmful materials; hence, it is a threat to groundwater sources.

Whatever the situation, these practices are being challenged in recent years due to the increasing value of land, inadequate space, and the limited carrying capacity of the environment; ultimately, they pose threats to human life.

The other face of waste is striking. Materials discarded as waste often contain huge amount of materials that are mined, refined, and rigorously processed to make final products for use by the end consumer. Enormous quantities of waste in various forms are released during this production phase as well. In all, waste, termed variously 'emissions', 'discharge', 'residue' and 'refuse', has become an inevitable component of modern life. Generation of waste and its management are not the only issues to deal with; depletion of natural resources is another problem. Humanity's ecological footprint has more than tripled since the 1960s. For the past 20 years, the earth's ability to support our lifestyle has been at stake, and we must halt the current trend. The need to balance our consumption and nature's capacity to regenerate and absorb our wastes is irrefutable.

In this context, managing the waste we generate and reducing the resources we consume have become a subject of immense importance. Reducing consumption, reusing materials and recycling products have proven to be practical solutions to the waste crisis and resource depletion concerns in many parts of the world. In recent years, reduce, reuse and recycle, the three Rs of waste management, have gained momentum in many developed countries.

Three Rs Initiative

Junichiro Koizumi, the former prime minister of Japan, proposed the Three Rs Initiative at the G8 summit held at Sea Island, Georgia, USA, in 2004, and it was formally endorsed by the leaders. Subsequently, a ministerial conference to launch the three Rs concept at a regional level was conducted in Tokyo in April 2005. The launch was considered the first step in changing global consumption and production patterns, and building a sound material-cycle society. Ever since, there have been significant developments on three Rs initiatives in most Asian countries.

Many non-governmental organisations (NGOs), community-based organisations (CBOs), and the private sector have been actively working across Asia to promote the three Rs in various forms, in a loose fashion, but coordinated, concerted efforts are needed to reap the benefits on a larger scale.

Three Rs endeavours in managing municipal waste

Managing municipal waste in urban centres is a challenging task, especially when the waste streams are heterogeneous with little or no collection fee. There are significant differences across Asia in the generation and composition of municipal waste due to varying geographic and climatic conditions, and lifestyles. At times, waste generation patterns differ even within the same municipality.

The composition of the MSW generated in some Asian countries has been presented in Table 1. The waste components are, in most cases, discarded or dumped without treatment or recycling. Such practices have prompted some private sectors and NGOs to initiate recycling and proper waste-management strategies. For instance, the Exnora & Pepsico Zero Waste Centre, an NGO based in Chennai, India, has been actively involved in promoting three Rs initiatives and good practices (Box 2).

Technology applications for thermal recovery (direct combustion of waste to recover heat) and fuel recovery are not found in most Asian countries. These technologies are applied only in the developed Asian countries.

In China and Thailand, these technologies do exist, but there is uncertainty over their efficiency in terms of both cost and environmental factors. Material recovery from, and sorting of, MSW remains largely unexplored in many Asian countries. Although some pilot models have proved successful in developed countries, further research is needed before they can be implemented.

In the developing countries of this region, a chain of informal recyclers – from waste scavengers to waste dealers – perform the task of material recovery and sorting. It is justifiable to state that their livelihood could be at risk if the technologies described above become operational and commercially successful. However, given the health risks and the need for resource conservation, the provision of these technologies – or at least some formal registration and support from governments – is vital. Attention should be paid to three Rs technologies associated with MSW, such as composting, which has proven to be one of the cheapest solutions for developing countries. Bangladesh has set an ideal example by successfully decentralising composting systems throughout the country. Similar approaches are also seen in China, India and Thailand, but they lack the fundamental support to make decentralised composting robust and successful.

Three Rs endeavours in managing e-waste

According to *The Financial Express* [8], about 80 per cent of the e-waste generated in the USA is exported to India, China and Pakistan. The recent ban on importing e-waste to China has diverted much of it to Bangladesh and other neighbouring countries due to the presence of cheap labour and recycling businesses. In Delhi alone, about 25,000 workers handle 10,000–20,000 tonnes of e-waste at scrapyards every year, with obsolete personal computers accounting for 25 per cent.

Box 2. Exnora and Pepsico Zero Waste Centre, Pammal, Chennai, India

Pammal, though administratively a town in the Kancheepuram district in the Indian state of Tamil Nadu, is a sprawling urban area of the Chennai City. With a population of more than 250,000 people, Pammal is also known for its world-class leather skins and hides. Today Pammal is an example of how civic engagement and people's partnership can be used to achieve source segregation of solid waste, vermi composting of organic waste, sale of recyclables and restoration of an ecosystem, the Pammal Lake, eventually leading to a Green Pammal.

Exnora Green Pammal is a unique joint venture of Private, Public and Government to improve the living environment and to promote more environment-friendly sustainable human settlements. Since its inception in 1994, Exnora Green Pammal has been able to standardize the collection and composting methodology over time to suit the local conditions.

The foremost priority of the Green Pammal Exnora was garbage removal and introduction of Zero Waste Management to cover the entire area. Zero Waste Management was achieved with a strategy of door-to-door campaigns and awareness-creation drives. The Zero Waste Centre has the following activities:

- producing premium compost from organic waste through Vermi composting;

- crushing and recycling plastic pet bottles;

- making recycled plastic products such as egg holder, hooks etc.;

- raising and selling ornamental and horticulture nursery seedlings;

- producing plates and cups of Areca leaves as substitutes for plastic plates and cups;

- selling other recyclable waste to respective industries.

Awards and Credentials

Golden Peacock Award 2006 for innovation in Waste Management

"Promising Practice" by UN-HABITAT, United Nation's Human Settlement Programme, 2006

UNICEF selected International Learning Center for 2007 under Urban Solid Waste Management

In China, intermediaries generally make e-waste imports possible. The waste is shipped to Hong Kong in containers labelled 'for recycling' and then smuggled to a number of recycling towns in China. Other Asian developing countries such as Sri Lanka, Nepal and Bangladesh have not been spared from this flow of e-waste. It is a crisis not only of quantity but also of toxic ingredients, such as lead, beryllium, mercury, cadmium, and brominated flame retardants, which pose both occupational and environmental health threats. But industry, government and consumers have, to date, taken only small steps to deal with this looming problem. Especially in developing countries, e-waste is highly sought after by scavengers and local recyclers.

Most of the developing Asian countries are at an early stage in implementing three Rs technologies related to electrical and electronic waste. The whole market is driven by chains of informal recyclers. Manual dismantling of electronic components is the most common method of recovering valuable materials. Electrical components (including wires) are burnt to recover copper and other metals. Other e-wastes are dismantled and sorted manually to recover components such as printed circuit boards, cathode ray tubes, cables, plastic, metals, condensers and batteries.

As in any other developing country, e-waste management is a major issue in Cambodia too. However, its economic status attracts more imports/dumped materials than local e-waste. Cambodia is starved of electricity supplies, and hence people in rural areas rely on batteries as a source of electricity. In such a situation, lead acid batteries play an important role in powering the economy. Various measures have been taken at the local and national level for the environmentally safe recycling and disposal of used lead acid batteries, which contain hazardous materials.

This growing concern over e-waste, domestically generated and imported, in the developing Asian countries has prompted many NGOs to initiate their own campaigns to promote the safe handling of e-waste. Current efforts in managing e-waste would be enhanced if policymakers and industries jointly implemented mechanisms such as 'extended producer responsibility'.

Three Rs endeavours in managing health-care waste

As mentioned in a World Bank report in 2006, management of health-care waste in the developing countries of Asia is not satisfactory. Uncontrolled burning, reuse of disposable items, and accidental injuries from improperly discarded needles are common and lead to life-threatening infections such as those of hepatitis B and C and HIV.

Even today, health-care waste management and its potential threat in most developing countries remains a subject that is not well defined and understood by either the general public or policymakers. As a result, health-care waste is often disposed of in general waste streams, although sometimes it is collected separately and burned in locally made or poorly maintained incinerators.

It is a mystery as to who is responsible for managing such waste once it reaches the dump or municipal collection point. (Open dumping is more prevalent in developing countries than waste disposal in landfills, even though the former is a designated area for waste disposal managed by municipalities.) Proper source separation would effectively answer such questions. It would also greatly reduce the amount and the toxicity of the medical waste requiring treatment and disposal.

For instance, the single-bin collection and storage system means that any waste generated within a medical facility is considered infectious. Three Rs initiatives could play a major role in reducing the amount of medical waste generated in the developing countries and in diverting such waste from municipal dumps.

To boost such initiatives and to solve the medical waste management dilemma, the current system needs a 'medical waste supervisor', as suggested by Dr Satoshi Imamura of the Japan Medical Association [9]. Such medical waste supervisors could play a significant role in managing medical waste and make a significant change in its handling and safe disposal in the developing countries.

Conclusion: three Rs for waste management is no longer optional but essential

The advantages of the three Rs in waste management are clear and need no explanation. However, the following propositions need to be considered in successfully implementing this:

- Zero-waste concepts, including three Rs, product stewardship, cleaner production and specification in the selection of packaging materials to manufacturers, needs to be promoted on a larger scale.

- Executable and realistic master plans and implementation plans for MSW management at the provincial level or state level in accordance with the national strategy should be made.

- The polluter-pays principle must be applied to all waste generators, especially in urban areas, including government agencies and departments. Countries such as Singapore and South Korea have them in place and have realised tangible benefits. Other developing countries need to adopt such systems as well.

- Technology transfer should be encouraged, but after testing the feasibility to suit local conditions and waste characteristics.

- Separate waste management systems for hospitals and other health-care establishments and industries should be created to prevent infectious and hazardous wastes from entering the municipal waste stream.

- Waste collection and disposal fees should be levied according to waste generation rates and the economic standard of the area, while considering the nature of the waste and resources needed for the treatment.

- Tax exemptions must be granted for importing recycling technology, and improved tax benefits for industries using waste and scraps as raw materials.

- Public education programmes, training and workshops, introducing solid waste management, and three Rs concepts in particular, should be encouraged.

- It is easier to mould young minds than old. Considering the resistance to change in adults, educating children about three Rs concepts should be done to have a better world in the future.

- Waste separation and recycling programmes for households, commercial centres, institutions and factories should be promoted.

Acknowledgements

The authors would like to express their deep gratitude to the Asian Development Bank and the Three Rs Knowledge Hub for making this chapter possible. All research work cited and quoted here is sincerely acknowledged.

References

[1]   *Toward Resource Efficient Economies in Asia and the Pacific – Highlights* (Asian Development Bank and Institute for Global Environmental Strategies, 2007).

[2]   Alan, B. 'Environmental Cost of Asia's Development', *Asia Times Online*, Hong Kong, 2002 (www.atimes.com/atimes/Asian_Economy/DK26Dk01.html).

[3]   *Down to Earth* (New Delhi: Centre for Science and Environment Publication, 15 March 2007).

[4]   Manandhar, R. 'Private Sector Participation in Solid Waste Management in Kathmandu', Kitakyushu Initiative Seminar on Solid Waste Management, 19–20 September 2002, Kitakyushu, Japan.

[5]   Visvanathan, C. and Trankler, J., Asian Society for Environmental Protection, Thailand. 'Municipal Solid Waste Management in Asia: A Comparative Analysis', *ASEP Newsletter*, 2003, 19(3); 20(4).

[6]   Kumar, S. 'Municipal Solid Waste Management in India: Present Practices and Future Challenge' in *Proceedings of the Hands-on Workshop on Sanitation and Wastewater Management*, ADB Headquarters, Manila, 19–20 September 2005.

[7]   Vidanaarachchi, C.K., Yuen, S.T.S. and Pilapitiya, S. 'Municipal Solid Waste Management in the Southern Province of Sri Lanka: Problems, Issues and Challenges', *Waste Management*, 2006, 26(8), pp.920–30.

[8]   'A Wiser Approach to e-Waste', *The Financial Express*, 2005 (www.financialexpress.com/fe_full_story.php?content_id=108565).

[9]   Imamura, S. 'Doctors' Efforts Toward Appropriate Medical Waste Management', Asia 3R Conference, 30 October – 1 November 2006, Tokyo, Japan.

9
Investing in waste-to-energy projects: success factors for public–private collaboration in Asia

Tim Forsyth

Development Studies Institute, London School of Economics and Political Science

This chapter discusses a key dimension of 'WASTEnomics' – that of turning the liabilities of agricultural waste in Asia into an asset and a valuable resource of energy through successful public–private collaboration for technology transfer in investing in waste-to-energy projects.

Introduction

Waste management in developing countries is both urgent and topical. At the local level, municipal waste is growing rapidly, and is the source of disease and pollution. Globally, it also contributes to anthropogenic climate change by releasing carbon dioxide if burnt, and methane through decomposing organic matter. Methane is an important greenhouse gas because it has 23 times the global warming potential of carbon dioxide.

Waste management is therefore an opportunity to address both local and global environmental problems, and hence it is a potential subject for the Kyoto Protocol's Clean Development Mechanism (CDM). The CDM was established to encourage investment in climate-friendly activities in developing countries in ways that also contribute to local sustainable development. Indeed, using waste to generate energy or electricity may contribute further to climate change policy by reducing the local need for fossil fuels, and encouraging some forms of renewable energy.

But waste-to-energy is also controversial. Burning agricultural waste to produce energy has long been practised in many places. Critics, however, suggest that municipal waste-to-energy offers different, and sometimes unacceptable risks. Simply incinerating waste, including plastics, may create hazardous ash and emissions such as dioxins. Moreover, some environmental non-governmental organisations (NGOs) claim that waste-to-energy technologies do not help overall environmental performance because they legitimise the creation of waste. Some critics are working to ban incineration of waste from the list of investments acceptable under the climate change negotiations.

Against these worries, some investors have argued that up-to-date technologies, such as pyrolysis,[1] provide effective alternatives to incineration if correctly applied. Furthermore, different technologies, notably biomethanation (or anaerobic digestion), extract methane from organic waste without burning, and offer opportunities for recycling inorganic waste. But these technologies offer additional challenges by being poorly understood in many locations, or are considered costly because they require careful segregation of organic and inorganic waste. Many companies fear communicating about these technologies, and building local trust will be a costly and long-term undertaking.

This chapter examines how these alternatives to incineration in waste-to-energy may be installed in developing countries by using collaboration or partnerships between investors and local citizens. It summarises the challenges for technology transfer for climate change mitigation, and how partnerships may assist. It then describes some examples of partnerships concerning waste-to-energy in Asia to draw practical lessons for potential investors in this field.

The problems of technology transfer

'Technology transfer' means increasing the adoption of a new technology by new users, or in a new region. Technology transfer is often discussed in climate change negotiations as an important way to achieve industrialisation without increased greenhouse gas emissions. But private companies have found it difficult to achieve for various reasons [1, 2]:

- There are frequently differences between how technology transfer is discussed in policy circles and in how investors actually operate. Rather than engaging in 'technology transfer', companies use 'leases', 'contracts' or 'joint ventures', all of which may include scope for encouraging technology use in new locations, but which are mainly aimed at increasing business access.

- Most environmental technology is now privately owned, and few companies wish to share it without compensation.

- Long-term technology transfer is costly, and requires training local people to use and maintain technologies; few companies wish to do this, and they often see it as the responsibility of international organisations or official development assistance.

---

1   Pyrolysis is a form of incineration that chemically decomposes organic materials by heat in the absence of oxygen. It typically occurs under pressure and at operating temperatures above 430 °C (800 °F).

- It is sometimes difficult to agree on what is 'environmental technology', and technologies have varying environmental effects for different stakeholders. Furthermore, many technologies developed in developing countries may also have environmental benefits and be more appropriate to end-users than some imports.

- Despite environmental benefits, some technologies have proved inappropriate for local users and have consequently been abandoned.

- Many programmes of technology transfer have failed to acknowledge the need for long-term financial security and cost-recovery by investing companies: companies require regular repayment of costs, and this may require the establishment of new accounting and financial bodies locally to achieve this.

- In addition, the common use of subsidies as an incentive to adopt new technologies has frequently backfired by creating short-term and unsustainable economic conditions that have repelled both investors and consumers.

Consequently, 'technology transfer' should not be seen as one simple process but rather the conjunction of various acts, over a long time, for a wide range of products and services, which have to be seen as *appropriate* by local users. Moreover, some analysts have suggested that technology transfer may follow two main paths. *Vertical technology transfer* involves the relocation (or sale) of technology products without the sharing of intellectual property, usually by the granting of sole production rights to one investor, or the simple sale of finished products to consumers in a new location. *Horizontal technology transfer* involves the long-term sharing of intellectual property, usually via a joint venture or cooperation between foreign direct investor and a domestic company in the host country.

Table 1 suggests some critical success factors underlying technology transfer as applied to renewable energy.

Table 1. Universal critical success factors for renewable energy development

1. Investment must fit the medium-term strategy of energy development.

2. Investment must use proven or reliable designs.

3. Projects must be based on least-cost approaches.

4. Appropriate finance must be arranged to cover risks.

5. There must be adequate marketing and technical staff.

6. There must be a proven market for the technology.

7. Do not give free gifts or overt subsidies (such as short-term grants).

8. Ensure that a market chain exists between suppliers and consumers.

9. Consider site-specific factors in each location.

10. Operate in locations where regulations and laws are favourable.

11. Create an acceptable tariff structure to cover costs.

12. Disseminate programme results to create market demand.

13. Conduct adequate project reviews to identify weak points.

14. Expect demand for products to grow once established.

Source: Forsyth [1], after Stainforth and Staunton [3].

Most discussions of technology transfer in international meetings to date have implied horizontal transfer, without much discussion of commercial fears of private companies.

For example, Chapter 34 of Agenda 21 (signed in 1992), urged developed countries to undertake technology transfer 'on favorable terms, including on concessional and preferential terms'. Some observers considered this practice to be commercially unsustainable. Similar exhortations were made in the text of the United Nations Framework Convention on Climate Change (UNFCCC) (also signed in 1992).

In 2000, a special report from the Intergovernmental Panel on Climate Change [2] identified technology transfer as a five-stage process, including assessment, agreement, implementation, evaluation and adjustment, and replication (diffusion) of both technology design and management/financial support. Other statements by the UNFCCC have reiterated the role of government action by listing activities such as providing information, financial flows, and improving legal frameworks. The Subsidiary Body for Scientific and Technological Advice (SBSTA) has been closely involved in developing a technology information system (TT:CLEAR[2]), including an inventory of the Energy Saving Trust (EST) and its projects.

But various observers have commented that most official attention to technology transfer has been given to state- and supply-led initiatives, and have suggested that socially concerned NGOs should be responsible for making contact with end-users of technology. In 2003, the UNFCCC asserted that 'governments can create enabling environments for EST diffusion and transfer if they endorse the importance of socially and environmentally oriented organisations and mandate social impact assessments for technology transfer projects' (p.16) [4]. But this view was challenged in part in 2004 by the Expert Group on Technology Transfer (EGTT)[3] of the UNFCCC,

---

2   http://ttclear.unfccc.int/ttclear/jsp/

3   The EGTT was established by parties at the Seventh Conference of the Parties to the UNFCCC (COP-7) held in Marrakesh in November 2001. The objective of the EGTT is to enhance the implementation of Article 4, paragraph 5, of the Convention such as by analysing and identifying ways to facilitate and advance technology transfer activities and making recommendations to the Subsidiary Body for Scientific and Technological Advice (SBSTA).

which instead urged that businesses should also be engaged more thoroughly in the diffusion of technologies, rather than seeking investors as responsible for innovation and development, with NGOs and governments as the only means of transfer [5]. But how can this be done?

The potential of partnerships

Partnerships or collaboration between investors and local citizens may allow technology investment to be more in touch with local needs, and to reduce costs of investment by sharing tasks with end-users. Collaboration may take place with citizen groups, local authorities, or small companies that are peopled by local citizens. This kind of partnership, or collaboration, is different in principle from existing forms of public–private partnerships between private companies and governments, such as Build-Operate-Transfer (BOT) schemes, because they are designed to build long-term trust and lower costs of investment, rather than simply provide environmental infrastructure by delegating investment to a private company.

Theorists have suggested that partnerships may benefit investors in two key ways: *transaction costs* and *assurance mechanisms* [6]. Transaction costs may be defined as costs of interaction (such as financial cost or time in negotiating with different actors), and assurance mechanisms may be defined as contracts, laws or expectations (formal or otherwise) that ensure that collaboration or partnerships provide each party with its desired result. An ideal partnership between actors should have minimum transaction costs, and maximum assurance mechanisms (Table 2). It should be noted, however, that the emergence of successful partnerships varies according to several factors, including willingness to cooperate, the long-standing trust of each party, and a shared or compatible perception of the underlying problem. Moreover, the ability to collaborate may vary among other companies, and among local citizen groups.

Partnerships may benefit communities by offering a form of 'cooperative environmental governance' [7], which refers to a system of decision-making about environmental technology and investment that includes the participation of local citizens, and the search for mutual objectives between investors and communities. Such partnerships are usually characterised by clear – and unanimously agreed – objectives of investment and technology, the existence of clear and accountable negotiating arenas where all citizens can express views, and, frequently, the existence of help from government departments (such as environmental agencies) in providing environmental and technical expertise.

Table 2. Conditions influencing the emergence and maintenance of collaboration

Transactions costs of alternative decisions

	High and applicable to all stakeholders	High for most, but not all stakeholders	Low
	1.	2.	3.
None	No collaboration	No collaboration	No collaboration
	4.	5.	6.
Partial	Collaboration possible, but not sustainable	Highly unlikely	No collaboration
	7.	8.	9.
Full	Sustained collaboration	Collaboration possible, but not sustainable	No collaboration

Source: Williamson [8]; Weber, p.34 [6].

It should be noted, however, that discussions of 'community' should not suggest that there are no differences between groups of people within localities or cities in developing countries. For example, some obvious differences between social groups involved in waste management include waste pickers – or people who make a livelihood by collecting waste that can be recycled – and richer, middle-class citizens, who may generate much of the waste. Moreover, it is also possible that partnerships involve some element of influence from the government or NGOs and activists outside localities. 'Local' cooperation, therefore, may be a misnomer. Indeed, some NGOs, such as Greenpeace, have in recent years opened offices in Asian cities and adopted international campaigns against toxic pollution, including waste-to-energy.

Some case studies: waste-to-energy investment in Asia

These case studies illustrate examples of waste-to-energy investment, using different technologies of incineration, pyrolysis and biomethanation in India, the Philippines and Thailand, in which partnerships between investors and local citizens were attempted.

Case study 1 – the importance of assurance mechanisms

Between 2000 and 2001, Enron, the US-based multinational energy investor, sought to develop a US$96m, 40-MW energy plant using rice husks in the province of Bulacan, in Luzon. Bulacan is one of the most important rice-growing regions of the Philippines, and the large quantity of rice husks produced as agricultural waste offered an important opportunity to use efficient incineration methods to convert them to energy. However, the project failed when the financiers learned how

Enron had organised its contracts for supplying rice husks. It had made contracts with some 150 rice millers to supply rice husks, and needed to maximise supply in order to fuel its large, 40-MW plant. The rice millers quickly discovered that Enron had no other suppliers of rice husks, and so could increase the price, thus eroding Enron's profitability. Under these conditions, the financiers withdrew their support.

An alternative outcome was illustrated by a different case in Thailand. Between 2000 and 2004, a Thai-owned company, AT Biopower, sought to build six, 16-MW power plants using rice husks in the central plains of Thailand. The plan differed from Enron's project in the Philippines in many ways. First, the Thai company sought to build a number of smaller power plants, rather than one large, 40-MW plant. Secondly, the investor used a variety of techniques to ensure that supply of rice husks remained constant – for example, making contracts with just 20–30 rice millers per power plant, rather than 150: and seeking to use just 10–15 per cent of their total rice husk production, rather than 100 per cent, as was the case in Bulacan. The power plants therefore experienced lower transaction costs through dealing with fewer rice millers than in the Philippines, and did not rely on each miller's total rice husk production. Furthermore, millers are contracted to produce a guaranteed quantity of husks: they are fined if they fail to deliver, yet are also rewarded with a yearly bonus if they achieve their target. All of these techniques are mechanisms to ensure that partnerships between companies succeed. Yet, they are also crucial to ensuring the successful embedding of new energy technologies.

Case study 2 – the importance of transaction costs

Transaction costs are the costs of interacting with partners, and usually refer to financial costs; time spent negotiating; and problems of misunderstanding. The best partnerships have fewest transaction costs. But defining transaction costs may also include knowing where to draw boundaries between partners, with regard to which activities each is responsible for. Examples from the Philippines show the need to reduce costs with different partners.

Between 1996 and 1998, a US-based investor in biomethanation sought to establish a new methane-recovery and electricity-generating plant in Ayala Alabang near Manila in the Philippines. The investing company used two techniques to reduce transaction costs and maximise revenue for itself. First it negotiated a contract with a local NGO to supply waste from pigs and cows in the region. This was in both parties' interests: the US investor did not want to spend money on collecting waste (it had no expertise in this area, and the transaction costs of paying local collectors were too high); in addition, the NGO wanted to reduce waste locally. Secondly, the NGO also negotiated another contract with the local municipal government to buy the entire municipal waste stream from the locality, and hired local waste pickers to sort it into organic and inorganic waste. Segregating the waste in this way is necessary in order to extract the organic material for biomethanation, and to make money from recycling inorganic material such as metal and paper.

Unfortunately, this investment project failed for several reasons. The most important was that local landowners (including the municipality) increased the rent payable on the power plant's land because they believed the project was more profitable than it was. But, in addition, the investing company quickly realised that the stream of recyclable (inorganic) waste was much smaller than it had anticipated because the waste pickers and waste transporters were removing the most valuable elements before they arrived at the plant. The company quickly decided that it could not control the supply of recyclable waste, and so it decided to omit waste recycling from its business objectives. It has since focused on biomethanation, composting and carbon credits as its main profits, and has left most recycling to the local people.

Using partners to reduce, rather than increase, transaction costs, seems to be the lesson. In other projects, local waste pickers have also been hired to collect or segregate waste because it allows investment projects to be seen by local people as opportunities rather than threats to their livelihood. It also allows investors to find areas of collaboration that maximise mutual benefits. The same US investor has later persisted with other biomethanation projects in the Philippines, notably in Baguio in Luzon, and General Santos in Mindanao, where local people are hired in order to undertake waste sorting, but where the investor does not seek to restrict the local people from recycling in ways that benefit them. Much of this success comes from defining boundaries around different business activities: the investor focuses on biomethanation and electricity generation, the local pickers on recycling. This way, both sides can maximise their own profits without undermining the partnership.

Case study 3 – the importance of trust and transparency

But partnerships between investors and local companies and citizens can easily be undermined by a loss of trust, or worries about the new technology. Local partnerships are not simply a pragmatic way of introducing new environmental technologies; they are also seen by many people as new business opportunities that benefit some people more than others, or as political acts. Often, the political perceptions of partnerships are controlled by factors outside the immediate control of investors. But what can be done to make partnerships acceptable?

In Thailand, AT Biopower (mentioned above) tried to build one 16-MW, rice-husk power plant in the central province of Suphan Buri in 2000. This time, the proposal caused widespread protests by local farmers, who feared the plant would extract water, reduce rainfall and cause pollution. There were even fears that the plant would cause sterilisation of anyone who walked under the power cables. Protests against the plant were reported in the national newspapers. These fears were caused by general worries about industrialisation and pollution from power plants in Thailand, and by (alleged) misinformation spread by people who wanted to influence where the plant would be located.

In the Philippines, investors in biomethanation have also experienced opposition from national and international NGOs opposed to waste-to-energy in general. Environmentalists (and especially the NGO Greenpeace) undertook a successful campaign to ban incineration of urban waste, and to enforce segregation of waste at source into organic and inorganic. These steps were taken to reduce the vast production of waste that is now overloading the Philippines' cities, and to resist incineration of waste. But this activism has also included opposition to biomethanation, even though it does not involve incineration. Few activists understand the process of electricity generation via anaerobic digestion, and some believe any form of waste-to-energy is unacceptable because it legitimises the production of waste. In the city of Baguio, in the northern island of Luzon, one US investor faced opposition from a local NGO who claimed that the biomethanation technology would destroy people's livelihoods by preventing them from making compost.

There are many examples of political activism undermining investment in new technologies. But how can companies overcome local resistance? In these examples, investors took several steps to improve local trust, and to seek win–win solutions.

In Suphan Buri, AT Biopower undertook an extensive public education campaign, seeking to explain how rice husks would lead to electricity generation without significant pollution. The investor also committed funds from the plant to support local community-development projects, and allowed citizens to monitor pollution, with a commitment to pay compensation if pollution exceeded limits.

In the biomethanation plants, the investors deliberately tried to win local support by offering jobs to the local waste pickers and other residents who were concerned. This approach was also attempted at a biomethanation plant using municipal waste in Lucknow, Uttar Pradesh. This plant, opened in 2003, was India's largest plant using biomethanation technology, and aimed to generate some 5 MW of electricity from 400–500 tonnes of organic municipal waste a day, operated by an Asian-based company with a variety of international shareholders. The company collaborated with an Indian NGO, Exnora, which is famous for involving waste pickers in urban waste management. Unfortunately, this plant was forced to close in late 2004 because it could not obtain a regular daily supply of waste. Despite the attempts at building trust, the assurance mechanisms were not there, and this seems to have been a mishandling by both the local state and the investors.

A further example of attempts to build trust occurred in the Indian city of Chennai (Madras). In 2000–04, an Australian investor sought to build a waste-to-energy plant based on pyrolysis of urban waste. This technology has received much criticism within India on the grounds that it may release too many pollutants (a claim the investor denies), and that it is an insufficiently short-term solution to the creation of urban waste. To conduct pyrolysis successfully, the company had to collect the entire municipal waste stream (including paper and plastic) and burn these to gain sufficient calorific values in the waste. Some recycling of items such as metals was possible, but there were fewer opportunities for local waste pickers to be involved. To justify this technological requirement, the investor deliberately argued that it was neither healthy nor fair for waste pickers to be involved in this way. But concerns about the impact and overall cost of this technology led the local government eventually to reject the scheme in 2004.

The implication of these examples is that governing public happiness by means of partnerships between local people and investors can be very difficult and beyond the control of investors. Most companies have tried to maximise public trust by providing information about the new technologies, and by including many different people in the production process. But some technologies – such as pyrolysis – must control more of the waste stream, and therefore provide fewer opportunities for local involvement. Furthermore, in the political battles surrounding the choice of waste-to-energy technology, statements are often not linked to localities, but come from national or international NGOs and activists.

Lessons for building partnerships for technology transfer

So, how can collaboration reduce the costs of investors, and increase the success of environmental technologies? The following five points seem clear.

(i) Be feasible

Most examples of successful partnerships are based on targets that are achievable and that can form a successful template for further projects. Enron's rice husk project in Bulacan in the Philippines, failed because it sought to generate 40 MW. But AT Biopower in Thailand has proven that smaller plants (of 16 MW) can work. Similarly, forming partnerships with fewer numbers of partners may be more achievable than with larger numbers.

(ii) Maximise assurance mechanisms

Assurance mechanisms are the devices – such as contracts and understandings – that keep both partners together in a partnership. In Thailand, AT Biopower successfully created incentives to ensure that the suppliers of rice husks honoured their contracts by making sure the power plant was not dependent on any one supplier, and by giving cash bonuses to suppliers who performed well. In the Philippines, investors in biomethanation sought successful collaboration with local citizens by ensuring that both parties had something to gain from the completion of power plants (i.e., citizens benefited from waste reduction and the opportunity to profit from recycling; the company gained from having access to the organic waste).

(iii) Minimise transaction costs

Transaction costs are the costs of interaction that can make or break a partnership. In the Philippines, investors in biomethanation realised that transaction costs would be reduced once clear boundaries were established around the ownership and participation in the waste treatment process. Successful assurance mechanisms can also mean reduced transaction costs, as both sides have incentives to perform. Reduced transaction costs usually mean understanding what aspects of the partnership are most important for one party, and specialising in these, rather than assuming that all aspects of interaction will be successful.

(iv) Be aware of politics

Political activism, and environmental campaigns may get in the way of successful collaboration. If companies establish successful assurance mechanisms and low transaction costs, there may be a small chance of political activism getting in the way of partnerships. But political activism may emerge for factors beyond companies' control, and may result from more general worries about the role of foreign investors in the domestic economy, the role of an allegedly corrupt local government in favouring one company above others, or fears about technology and environment that may or may not be well founded. In such cases, some companies have responded by trying to control their own image. The Australian investor in Chennai, for example, tried to legitimise pyrolysis by de-legitimising using waste pickers in waste management. In the Philippines, some activists unfairly accused biomethanation of being another form of incineration. In these cases, companies have responded by engaging in gentle dialogue with critics, and by including some element of community development in their projects. In many ways, being aware of politics is a broader way of maximising assurance mechanisms and minimising transaction costs.

(v) Work with others

Finally, partnerships often result not only from the hard work of specific companies or business managers, but from the coincidence of various local, national and international factors. In the Philippines, a national law requiring all municipal waste to be segregated may help partnerships emerge between investors who want to build biomethanation plants and local citizens who are worried about increasing waste totals. In Lucknow, India, the local government assisted the new biomethanation power plant by urging companies to adopt a positive attitude to hiring local waste pickers (although it could not supply the waste). National and local NGOs may also seek to engage constructively with companies – for example, the NGO Exnora in India has established beneficial relationships with some waste management companies. For environmental policy, involving different actors in business, society, and government increases the chances of cooperation, and decreases dependency on any one actor. Table 3 summarises some potential roles played by different actors.

In turn, these lessons also illustrate the following two general lessons for partnerships in technology transfer for climate change mitigation.

First, capacity building for technology transfer should not be simply defined as 'transferring experience, knowledge, skills and practices' (p.4 [9]). Rather, capacity building should include focus on strengthening existing factors that allow actors to reach agreement. In essence, this requires seeking ways for investing companies to recover costs on a long-term basis, and appreciating that local users need to see technologies as appropriate.

Secondly, policymakers should acknowledge that 'communities' are more diverse than commonly described. Partnerships, or collaboration, in technology transfer should not be described in romantic terms as engaging with entire communities, as it will inevitably create winners and losers within communities. Seeking contractual arrangements or commercial partnerships between communities and investors will rarely involve all citizens. Recognising the diversity of needs and actors within communities may help capacity building for technology transfer by identifying different opportunities for appropriate technology.

Partnerships have the potential to overcome some of the barriers of costs and learning associated with new technologies. Looking at them as arenas that can reassure both investors and end-users may both offer important ways to enhance waste management in developing countries and contribute to global climate change mitigation.

Table 3. Building capacity for climate technology transfer via partnerships between communities and investors

Actions for national governments

- National legislation such as the Philippines' Clean Air Act and Solid Waste Act, which seek to attract investment in 'clean' technologies, educate residents about waste segregation, and prepare waste for treatment.

- National programmes for building investment in renewable energy technologies, such as Thailand's Small Producer Programme and Biomass Programme, which offer an initial subsidy for plants to invest in new technologies to use waste products for electricity generation

Actions for local governments

- Seek strong action and united support for projects that integrate waste management with generation of electricity.

- Seek support from national or local NGOs to ensure that any investment does not result in costly disputes.

- Ensure that benefits of new technology schemes are seen to be distributed locally, such as access to the electricity generated or by-products of waste segregation.

Actions for businesses and investors

- Seek collaboration with local NGOs or citizen groups who may be able to point to synergies and complementarities in aims that may lead to cost-saving opportunities.

- Allow time and money for educating residents about the objectives of the investment and technology, including frank discussion about who wins and loses.

- Avoid depending on a limited number of suppliers or collaborators, as they may be willing to exploit this dependency later on.

Actions for citizen groups and NGOs

- Seek collaboration with businesses with which there may be complementary aims, as they may provide commercial incentives for public-policy objectives such as waste collection, or training of unskilled workers.

- Participate in training and education if possible.

Actions for all actors

- Seek public debate about public–private collaboration, how private and public objectives may offer complementarity, how past experience may shape current perceptions of collaboration, and how cooperation may benefit all parties if conducted in acceptable ways.

References

[1]   Forsyth, T. *International Investment and Climate Change: Energy Technologies for Developing Countries* (London: Earthscan, Royal Institute of International Affairs, 1999).

[2]   IPCC (Intergovernmental Panel on Climate Change). *Methodological and Technological Issues in Technology Transfer: Special Report of Working Group III* (Cambridge: Cambridge University Press, 2000).

[3]   Stainforth, S. and Staunton, S. 'Critical Success Factors for Renewable Energy Development in Asia'. Unpublished paper, ETSU, Oxford, Government of UK, 1996.

[4]   UNFCCC. 'Enabling Environments for Technology Transfer', United Nations Framework Convention on Climate Change, FCCC/TP/2003/2, Bonn, 2003.

[5]   UNFCCC. 'Summary of the Senior-Level Round-Table Discussion on Enabling Environments for Technology Transfer, Held at the Ninth Session of the Conference of the Parties', United Nations Framework Convention on Climate Change Subsidiary Body for Scientific and Technological Advice, Twentieth Session, Bonn, 16–22 June 2004, FCCC/SBSTA/2004/2.

[6]   Weber, E. *Pluralism by the Rules: Conflict and Co-operation in Environmental Regulation* (Washington, DC: Georgetown University Press, 1998).

[7]   Glasbergen, P., ed. *Co-operative Environmental Management: Public–Private Agreements as a Policy Strategy* (Dordrecht/London: Kluwer, 1998).

[8]   Williamson, O. *The Mechanisms of Governance* (Oxford: Oxford University Press, 1996).

[9]   UNFCCC. 'Capacity-Building in the Development and Transfer of Technologies'. Technical paper, United Nations Framework Convention on Climate Change, FCCC/TP/2003/1, Bonn, 2003.

10
Waste-to-energy in Europe: environmental, energy-efficiency and financial considerations

Ella Stengler
CEWEP

This chapter discusses a key dimension of 'WASTEnomics' – that of turning the liabilities of waste into assets by recovering energy from waste.

Waste-to-energy plants are an essential part of waste management and the energy-supply network in Europe. They generate electricity and heat through the thermal treatment of municipal waste that is not reused or recycled by other means. Waste-to-energy contributes to climate protection and is a reliable method in waste management and energy policy. It goes hand in hand with recycling and is instrumental in reducing dependence both on landfilling and on fossil fuels.

Introduction

At present in the European Union (EU), municipal waste is disposed of through landfill (45 per cent), waste-to-energy (18 per cent), and recycling and composting (37 per cent).

Despite the EU policy of diverting biodegradable waste from landfill, landfilling remains the main method in Europe. There are many reasons for this, including the fact that landfilling is the cheapest way to dispose of waste in many countries and that investment costs for recycling and waste-to-energy plants are quite high. Such plants need planning security and rely heavily on waste-management policy (to divert waste from landfills). The public distrust of waste-to-energy plants also plays an important role, despite the fact that these plants now achieve very low emission levels.

Waste-to-energy plants generate energy from the waste and deliver it to homes and industry. They play an essential role in the European energy-supply network and contribute to ensuring that supply. In some larger Swedish cities, for example, waste-to-energy plants provide up to 50 per cent of the total heat demand.

Currently, about 58.5 million tonnes of municipal waste are thermally treated each year in about 418 waste-to-energy plants in Europe. The distribution of these plants is shown in Figure 1.

Figure 1. Waste-to-energy plants in Europe operating in 2005. Source: CEWEP

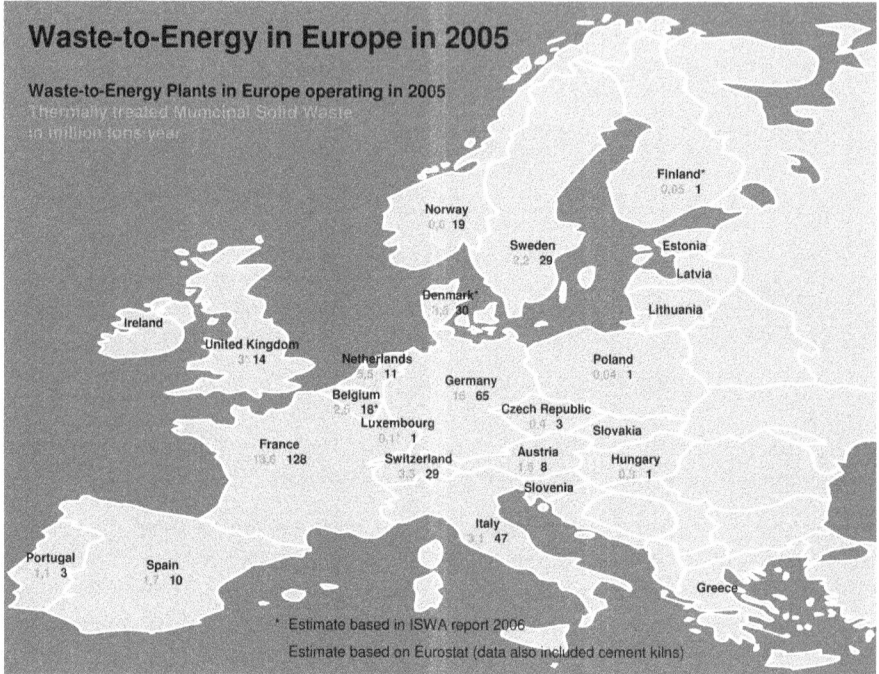

Waste-to-energy and the environment

Low emissions

In September 2005, a report [1] by the German Environment Ministry (BMU) stated:

- Because of stringent regulations waste incineration plants are no longer significant in terms of emissions of dioxins, dust and heavy metals. And this still applies even though waste incineration capacity has almost doubled since 1985.

- Total dioxin emissions from all 66 waste incineration plants in Germany has dropped to approx. one thousandth as a consequence of the installation of filter units stipulated by statutory law: from 400 grams to less than 0.5 grams.

In reference to other industries, the report says:

- The decline, however, has nowhere been as drastic as in the incineration of household waste. The consequence is that whereas in 1990 one third of all dioxin emissions in Germany came from waste incineration plants, for the year 2000 the figure was less than 1%.

These comments reflect the improvements that have occurred in reducing emissions in Germany, as a result of strict national legislation and significant investment in efficient flue gas-cleaning systems.

At the European level, the Waste Incineration Directive 2000/76/EC introduced strict emission limits – much more stringent than for any other industrial activity. More information on emissions from waste-to-energy plants can be found in the Best Available Techniques Reference (BREF) document for waste incineration [2].

Sustainable energy from waste

Waste-to-energy plants produce heat and electricity from waste, delivering it to households and industry, thus replacing the energy generated by conventional power plants using fossil fuels. Since about 58.5 million tonnes of municipal waste is treated annually in waste-to-energy plants across Europe, 23.4 billion kWh of electricity and 58.5 billion kWh of heat are generated each year. In effect, 6–32 million tonnes of fossil fuels (gas, oil, hard coal and lignite) can be replaced annually, which would have emitted 16–32 million tonnes of CO_2.

Replacing these fossil fuels, waste-to-energy plants can supply annually about 7 million households with electricity and 13.4 million households with heat (Figure 2). This is equivalent to supplying the entire populations of Portugal, Estonia and Denmark with electricity and the entire populations of Belgium, Hungary, Bulgaria and Norway with heat from waste-to-energy plants throughout the year.

According to the European directive on the promotion of electricity produced from renewable energy sources (2001/77/EC) (RES Electricity Directive), the biodegradable fraction of waste is considered biomass and is thus a renewable energy source. The biodegradable fraction in municipal solid waste is more than 50 per cent; according to a study by the Öko-Institut, it is 62 per cent [3].

In practice, member states differ in how they support electricity from waste and how they implement the RES electricity directive. While a number of member states recognise waste as a renewable energy source, only a few really support it. Thus, the price that waste-to-energy plant operators receive for selling their electricity ranges from the market price of around 4–5 eurocents/kWh in Germany to 10 eurocents/kWh; this can be achieved with green certificates in the Flemish part of Belgium. In Hungary and Portugal, operators get about 7 eurocents/kWh. Grid access plays an important role in supporting alternative energy sources.

There is considerable potential for waste-to-energy plants to contribute to climate protection through generating energy. But while the RES electricity directive applies this idea to electricity, the heat sector is not covered by EU legislation. This should be considered in future EU legislation. The European Commission (EC)'s new energy package includes a renewable energy directive and is to tackle heating and cooling from renewable energy sources.

Figure 2. Sustainable energy from waste

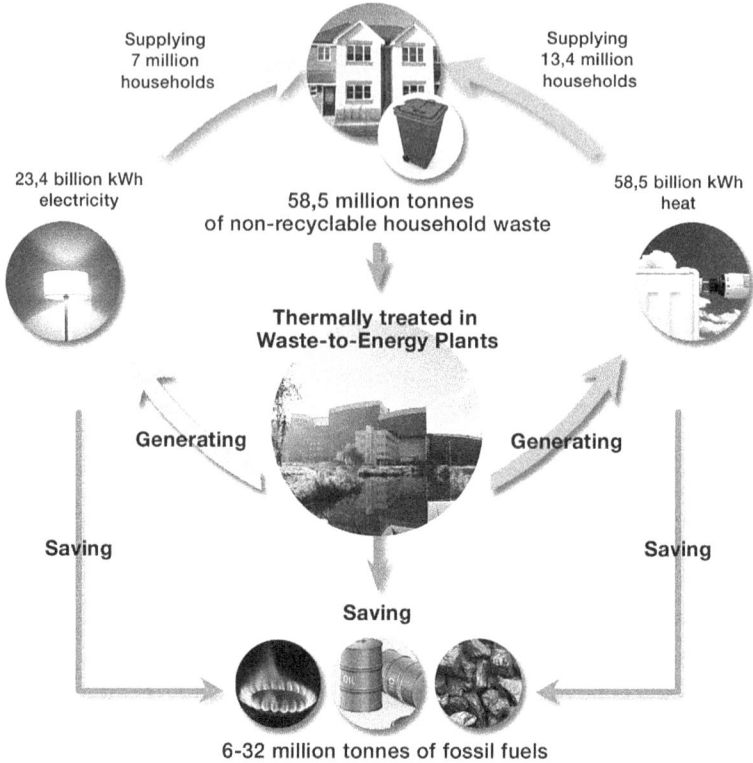

Supplying
7 million
households

Supplying
13,4 million
households

23,4 billion kWh
electricity

58,5 million tonnes
of non-recyclable household waste

58,5 billion kWh
heat

**Thermally treated in
Waste-to-Energy Plants**

Generating

Generating

Saving

Saving

Saving

6-32 million tonnes of fossil fuels

Impact of forthcoming EU legislation

On 21 December 2005, the EC delivered a proposal on the review of the Waste Framework Directive (WFD) and a communication on a thematic strategy on waste prevention and recycling. The WFD will give approximately a 20-year timeframe for European waste policy and will serve as a basis on which the waste management industry has to work.

The EC is seeking to clarify the definition of recovery and disposal. It proposes that the principle of 'replacement of resources' is still decisive in the definition of recovery. Recovery operations must serve a useful purpose in replacing, 'whether in the plant or in the wider economy', other resources that would have been used to serve that function. Substitution is now assessed from a wider perspective than the rigid approach taken by the European Court of Justice [4], which only considered the boundaries of the plant itself and stigmatised a municipal waste incineration plant as a waste-disposal operation.

Clarifying the energy-recovery status within the WFD is important for waste-to-energy plants; otherwise, they would be consigned to the bottom of the waste hierarchy, at the same level as landfills.

The definition which is now proposed by the EC in principle allows waste-to-energy plants to be classified as energy-recovery facilities, if the energy efficiency criteria described in Annex II (R1 formula) are met. Meanwhile the European Council agreed on this approach in its common position, adopted on 20 December 2007. At the same time, it extended the power of member states to apply the principles of self-sufficiency and proximity to shipment of waste destined to incinerators that are classified as recovery in certain specified cases (Article 14).

In February 2007, the European Parliament voted on the first reading of the WFD. They voted against the R1 formula for various reasons. Some MEPs (Members of the European Parliament) opposed the formula because they do not like the recovery status of waste-to-energy in principle, while others opposed it because they felt that the factors chosen by the EC are too high.

Although MEPs deleted the formula from Annex II, the definition they chose for energy recovery would mean that waste-to-energy plants could gain energy-recovery status. In the words of Art. 3 point (ib) (new), amendment 26,

- 'energy recovery' means the use of combustible waste as a fuel for generating energy through direct incineration with or without other waste or fuel but with recovery of the heat. Incineration of waste where more energy is added than received during the process is not treated as energy recovery.

The WFD returns to the European Parliament for a second reading in early 2008.

To coincide with the discussion on energy recovery, a study calculated waste-to-energy's contribution to climate protection [5]. This study shows the possible positive effect the energy efficiency formula proposed by the EC in Annex II, point R1 of the WFD, could have on CO_2 emissions. The study, based on the IPCC (International Panel on Climate Change) method calculated that, by the energy efficiency formula, up to 45 million tonnes of CO_2 emissions per year could be saved additionally. This is equivalent to the annual emissions of 20 million cars.

This could happen if efficient waste-to-energy plants achieve the recovery status, as they are higher in the waste hierarchy than landfills. The study is based on the assumption that an ambitious policy in the EU aiming at 60 per cent recycling and efficient waste-prevention measures will be implemented by 2020. The remaining 40 per cent of the waste should be used to generate energy.

Investment costs for waste-to-energy plants

The investment costs differ widely from country to country, for reasons such as employment costs, but the prices for materials also tend to be very different. The costs depend on the size of the plant, but also on the price of the land and on the equipment, such as additional facilities for sorting or treatment of slag. The cost of the flue gas-cleaning system, which guarantees the low emissions, accounts for up to almost 50 per cent of the total investment cost (Figure 3).

Figure 3. The waste-to-energy plant in Würzburg – the right hand section is the flue gas-cleaning system

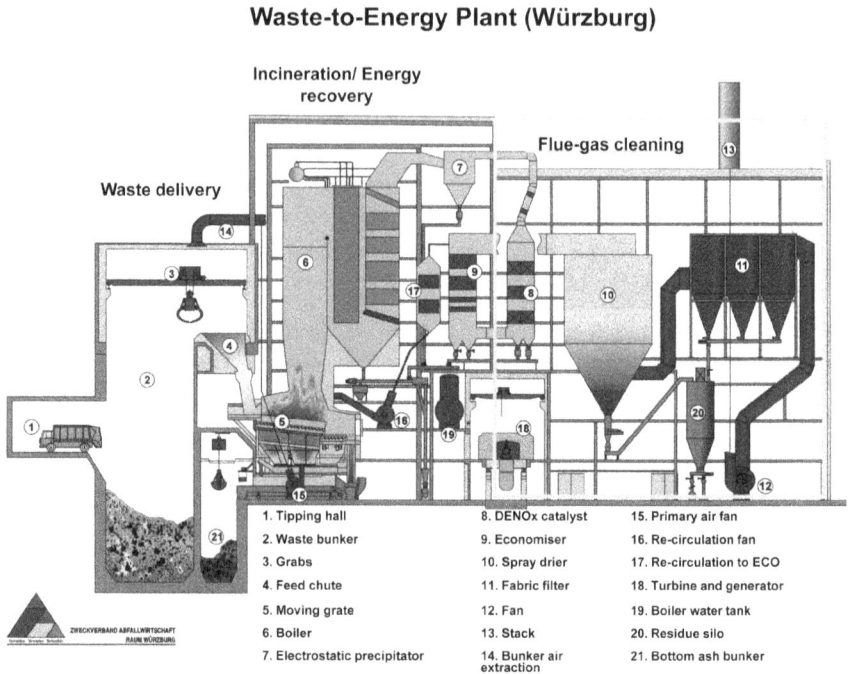

Waste-to-Energy Plant (Würzburg)

1. Tipping hall	8. DENOx catalyst	15. Primary air fan
2. Waste bunker	9. Economiser	16. Re-circulation fan
3. Grabs	10. Spray drier	17. Re-circulation to ECO
4. Feed chute	11. Fabric filter	18. Turbine and generator
5. Moving grate	12. Fan	19. Boiler water tank
6. Boiler	13. Stack	20. Residue silo
7. Electrostatic precipitator	14. Bunker air extraction	21. Bottom ash bunker

Therefore, it is very difficult to estimate investment costs. The two following examples of recently constructed waste-to-energy plants give a rough picture.

Example from Germany

TREA Breisgau has been in operation since March 2005. It has a four-phase flue gas-cleaning system. Its energy production is about 15 MW of electricity (supplying about 25,000 households) and about 20 MW of district heating (supplying about 1,500 households). Its capacity is 150,000 tonnes/year, and its investment cost is 83 million euros (= 533 euros/tonne capacity).

Example from Ireland (County Meath)

The investment costs for 150,000 tonnes/year capacity will be approximately 85 million euros (= 567 euros/tonne capacity). However, this includes material-recycling facilities with an annual capacity of 20,000 tonnes. The planned waste-to-energy plant will generate 13 MW of electricity, enough for over 19,000 homes annually.

In many European countries, the municipalities are responsible for household waste and the private sector for industrial waste. Waste-to-energy plants generally have contracts with the municipalities in order to guarantee long-term, environmentally sound waste treatment. Sometimes, the municipalities construct, often in cooperation with other municipalities, waste-to-energy plants and operate them on their own. Quite often, they leave the operation to a private company (public–private partnership). The gate fee for 1 tonne of municipal solid waste differs across Europe, in the range 30–250 euros/tonne (with a downward tendency). The gate fee for the waste treatment can be reduced if the revenue from the energy sold is higher. However, this differs widely, depending on the ability of waste-to-energy operators to deliver their energy to homes and industry, and particularly on access to the grid.

Comparing costs, a study carried out by ia GmbH on behalf of CEWEP, shows clear advantages of the thermal treatment of waste through a life-cycle comparison between a waste-to-energy plant and a landfill site in Europe by the method of total cost of ownership (TOC). This is particularly so if the aftercare period for landfills is taken into account [6].

Conclusions and outlook

For the waste-to-energy sector, it is vitally important to be classified as an energy-recovery operation within the review of the WFD.

Since the waste hierarchy gives priority to recovery rather than disposal, it would be counter-productive for European waste policy that aims to divert waste from landfills if waste-to-energy plants were to have the same classification as landfills, that is, disposal, considering that landfill gases, which are mainly methane, contribute more significantly to global warming than the net emissions of CO_2 from waste-to-energy plants.

It should be mentioned that any fears that waste-to-energy might inhibit recycling are not justified. In fact, the two treatment options go hand in hand. Countries whose recycling levels are among the highest in Europe, such as The Netherlands, Germany, Sweden and Switzerland, also have high levels of waste-to-energy. Low waste-to-energy rates are synonymous with low recycling levels [7]. Through capacity planning, any encroachment into recycling markets can be avoided. Waste-to-energy and recycling are complementary options to reach the goal of European environment policy of diverting waste away from landfills.

According to the Landfill Directive 1999/31/EC, the amount of biodegradable waste going to landfills must be reduced to 35 per cent of the total amount (by weight, referring to the amount in 1995) by 2016. As a result, implementing this directive will reduce greenhouse gas emissions by around 74 million tonnes in CO_2-equivalents [8]. In order to reach this goal, a greater capacity of both recycling and waste-to-energy is needed. Investment in new waste-to-energy plants, which is particularly needed in EU member states that still mainly rely on landfilling, would be easier under energy-recovery status.

Stigmatising waste-to-energy plants as disposal operations would result in steering the waste into industrial plants such as cement kilns, which are regarded as energy-recovery options by the European Court of Justice [9]. However, this is counter-productive for the environment, as waste-to-energy plants are equipped with efficient (though more expensive) filtration devices in order to achieve very low emissions. Therefore, there is no level playing field, and the market is distorted if waste-to-energy plants are classified as disposal operations, while industrial plants that co-incinerate waste qualify as an energy-recovery option.

Criteria for investors/government

Confidence in national waste-management policy

Investment in high-tech waste-to-energy plants needs planning security. This means reliance and confidence in the government's waste policy, as in capacity planning. If the government is not seriously trying to divert waste from landfills, investment in high-tech waste-to-energy plants is too risky. If government policy is to ship the waste abroad, investment in waste-to-energy plants would also be very difficult to realise.

An example is the decision by the Swiss government to ban landfilling of non-pre-treated waste. Thorough capacity planning took place, while policy was strict and reliable. So, step by step, the necessary waste-to-energy capacity was implemented, while at the same time landfilling was continually reduced. Waste-to-energy plants are well accepted by the public in Switzerland. There are some critical voices, but the discussion is not dogmatic.

Public acceptance and communication

Transparent information to the public is necessary for the acceptance of waste-to-energy plants. Governments and investors have to inform the public thoroughly about emissions, energy production, alternative waste treatment options, etc. Waste-to-energy plants have a very low environmental load per kWh, even compared with other sustainable energy sources. This insight is generally a determining factor in decisions concerning energy and waste.

Location and energy efficiency

The energy efficiency of a waste-to-energy plant is of the utmost importance, particularly considering the rising oil prices and Europe's dependence on fossil fuels. To obtain optimal energy efficiency, it is important to place a new waste-to-energy plant close to a potential user of the heat that will be delivered by the waste-to-energy plant, because heat cannot be transported over long distances. Good grid access is also very important for energy supply. Scandinavian countries, for example, have a very good infrastructure. Government policy on supporting renewable energy is also important.

Recognition of waste-to-energy plants as an energy-recovery option

Status as an energy-recovery facility can be decisive, especially to receive investment. Classifying a waste-to-energy plant as a waste-disposal facility will only discourage potential investors. Should waste-to-energy plants be classified as disposal operations, there is a distortion of competition in comparison with industrial plants co-incinerating waste and with (mostly cheaper) landfills. Therefore waste-to-energy plants should be recognised as energy-recovery facilities.

Financial incentives

Besides the implementation of a landfill ban, as in Denmark, Sweden, Switzerland and Germany, landfill taxes can also give an important incentive to divert waste from landfills to waste-to-energy plants. Currently, the highest landfill tax has been introduced in The Netherlands, a tax of 89.91 euros/tonne [10] on combustible waste. As a result, only 2 per cent of municipal solid waste is landfilled in The Netherlands.

References

[1] Federal Ministry for the Environment, Nature Conservation and Nuclear Safety. *Waste Incineration – A Potential Danger? Bidding Farewell to Dioxin Spouting.* September 2005 (www.bmu.de/english/waste_management/downloads/doc/35950.php).
[2] European IPPC Bureau. 'Integrated Pollution Prevention and Control', Reference Document on the Best Available Techniques for Waste Incineration, July 2005 (http://eippcb.jrc.es/pages/FActivities.htm).
[3] *Waste Incineration: Useful for Climate Protection.* Study carried out by the Öko-Institut Darmstadt on behalf of ITAD, the German waste-to-energy association. An English summary prepared by CEWEP is available on its website (www.cewep.eu/climateprotection/studies/art157,62.html).
[4] Judgment of the Court (Fifth Chamber) of 13 February 2003 in Case C-458/00: *Commission* v *Grand Duchy of Luxembourg* (www.curia.eu.int).
[5] *Opportunities to Reduce Climate Change by Using Energy from Waste.* Study carried out by FFact (www.cewep.eu/studies/climate-protection/art230,236.html).
[6] *Life Cycle Comparison of a Waste-to-Energy Plant and a Landfill Site in Europe by the Method of Total Cost of Ownership (TCO).* ia GmbH, August 2006 (www.cewep.eu/studies/costs/index.html).

[7]   'Don't Waste Waste – It Is a Resource' (www.cewep.eu/home/news/art150,232.html).

[8]   'Environment Research Plan of the German Federal Ministry for the Environment, Nature Conservation and Nuclear Safety', August 2005 (www.bmu.de/files/pdfs/allgemein/application/pdf/klima_abfall_en.pdf).

[9]   Judgment delivered on 13 February 2003, C-228/00: *Commission* v *Germany* (www.curia.eu.int).

[10]   An overview on landfill bans and taxes in European countries is available on CEWEP's website (www.cewep.eu) [click on data].

11

Applying industrial waste management in practice – reassessing the economics of the waste hierarchy

Martin Kurdve

Volvo Technology, Sweden

This chapter discusses a key dimension of 'WASTEnomics' – that of turning the liabilities of industrial waste into assets.

Industrial waste is becoming more important as a major cost, but also as a potential resource, in production applications. This chapter, developed for waste managers in the Volvo Group, shows how progressive work with the five-step approach to dealing with various waste streams can be environmentally sound and economically beneficial. It also stresses the importance of setting long-term goals in order to maintain competitive industrial advantages, and monitoring the performance by measurable indexes. In order to achieve this, a good knowledge and control of where waste streams are generated is needed. The aim is long-term control over waste generation and the level of waste toxicity as well as determining which waste contributes to the highest cost.

Introduction

Waste management is becoming increasingly important due to increasing legal requirements and the rising costs of waste disposal. The focus of both government authorities and companies has shifted from mere disposal to integrated management of resources and residues [1]. Due to the price of raw material in pre-industrial societies, it was often profitable for individuals to reuse scrap materials. Today, with mass production and mass consumption, the feasibility of this approach is less apparent, but, for society at large, due to both the scarcity of raw materials and the need to reduce emissions, integrated management of resources and residues is very important. This chapter presents general guidelines on efficient and sustainable handling of waste and rest materials.

This chapter is an overview of why and how we structure work in minimising waste volumes and using waste as a resource. It is not a description of how to work with every waste fraction, but it gives some examples of management of common waste types. An introduction is presented on how to prioritise the management of different waste types and different ways of processing rest material and how to take control of waste flows and the associated cost. Finally, some useful key performance indexes (KPIs) and follow-up of waste-reduction goals are suggested.

Waste management in the Volvo Group

Today, continual improvement of waste handling takes place at most Volvo Group industrial sites. However, as there is little contact between the sites, there may often be less than optimal waste management at the company level. There is also a recognised need to bring the focus of waste management upstream – closer to the stage where waste is generated. The environmental council at the Volvo Group has demanded a stronger focus on the handling of waste and rest material in order to find synergies between the plants and to reduce costs and associated environmental effects. The annual cost of waste within the Volvo Group increased by more than 35 per cent in 2004. This is substantially more than the increase of around 3.5 per cent in total waste volume for the years 2004–06, according to the environmental report. The cost increase stemmed from regulation issues, taxes and rising prices from waste contractors. From an environmental point of view, proper waste management helps to save limited resources, reduce landfills, and lower global warming potential (GWP). The potential saving by reducing waste is very large. According to studies in the 1990s by the US National Academy of Engineering, more than 90 per cent of materials used never end up in saleable products [2].

Box 1. Why is good waste management important?

- Minimising environmental pollution helps to protect human, animal and plant health.

- Minimising environmental pollution helps to save money.

- Reuse and recycling may generate revenue.

- It complies with regulations.

Source: Volvo Trucks Corporation [3]; Volvo Trucks waste guidelines for dealers.

A hierarchy for managing waste

What is waste?

According to the European Union (EU) and Swedish legal definition of waste, 'waste shall mean any substance or object included in one of the waste categories and which the holder discards or intends or is required to discard' [4]; that is, all material not part of the main output of an enterprise is regarded as waste.

This wide definition of waste may be a problem when it comes to legislation on various waste materials. In these waste management guidelines, however, it is a useful approach since waste management has to consider all products and material that may become waste. This means that, according to the EU definition, waste management should consider not only material with no value for the owner but also by-products and even some raw material.

Five-step stairway to heaven

The well-known five-step model can be used to analyse waste streams from an environmental point of view (Figure 1). This five-step approach is in line with the intentions of the EU legislation and with the EU waste hierarchy. The hierarchy implies that the priority should be to reduce the amount and toxicity of the waste and residual material. The waste that still arises should be reused or recycled, and resources should be recovered. Landfills should be considered the least desirable measure [5].

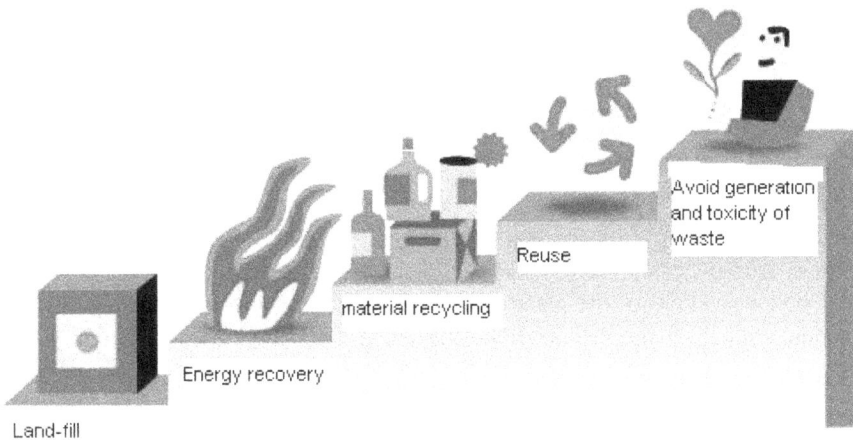

Figure 1. Five-step waste model (Picture from IL-recycling presentation material, 2006).

The five-step approach stipulates the following prioritisation of issues in waste management:

1. *Avoid* generation and *reduce* toxicity of waste; this can be facilitated by increasing the lifetime of products or by simply not using the products that end up as waste.

2. *Reuse* the products; after sorting and cleaning, many products may be reused. In some closed-loop systems, this is the same as increasing product lifetime as above. In other cases, products are reconditioned on- or off-site and then reused, sometimes as secondary products with less quality demand (as in the example of reusing cutting oil in Box 2).

3. *Material recycling.* If different materials are kept separated, they can be recycled as a raw material for new products. In some cases, sorting of the waste material is needed before material recycling can be considered. Some materials are reused mainly as secondary filling material; obviously, this may not be as 'good' as recovery of the original raw material.

4. *Energy recovery.* This is usually a euphemistic term for incineration even though it may also be gasification into a fuel. Incineration does recover some energy from material, and it lowers the amount of waste that is dumped in landfills. From 5 to 30 per cent of the original weight of most materials becomes ash. Sometimes the ash is used as filling materials for construction; therefore, this could be considered a partial material recovery. However, most of the ash is dumped in landfills or is spread, causing air contamination.

5. A *landfill* is the last resort for waste. It should receive only materials that cannot be otherwise disposed of, such as hazardous or contaminated waste.

From the long-term economic and environmental view, the first step, avoid generation of waste, is the best. For each stream, analysis should start by considering whether the waste can be avoided, how much it would cost to change, and how much time would be needed to make avoidance economically feasible. This step is the one that demands the greatest creativity, and 'rethink' is a key word often used together with 'reduce' and 'avoid'.

'Reuse' is the second-best approach from the perspective of long-term economic and resource efficiency. Reuse often involves cleaning and sorting. These processes may be in-house solutions at the plant, or the products may be sent to the supplier for reconditioning. As mentioned earlier, at Volvo, the cost of waste has increased rapidly over recent years; therefore, not losing material as waste results in economic savings.

Material recycling is an option used for products that cannot be mended or cleaned and sorted well enough for reuse. It usually involves a raw materials manufacturer that takes used instead of new material and produces new products from it. When the replaced material is less valuable than the original material (or the replaced product in reuse has less value than the original), the process is called down-cycling. An example is the use of plastic as filling material in construction.

Material that cannot be recycled can often be incinerated to recover some of its energy and minimise the volume of waste. In northern Europe, these incineration plants are often used for community heat supply (non-hazardous waste) or for cement kilns (hazardous waste) where heat is recovered and ashes are bound into the cement.

Finally, landfill disposal should normally be avoided; some inert rests or inseparable material cannot be processed by any of the other schemes and must be deposited in landfills. This is increasingly expensive and thus additional separation techniques will be used to minimise this type of waste.

Box 2. Reuse of cutting oil

One Volvo plant has successfully reused cutting oil. First, operations are adjusted so that the same type of oil is used for most operations. The cutting oil is recovered from the metal chips and cleaned for reuse in the same plant. A few operations are sensitive to variations in lubrication and thus only use fresh oil. The less sensitive operations, however, use all the recovered oil and some use new oil. The result is a reduction in overall oil use of around 50 per cent with an approximate 30 per cent cost reduction.

Real-life waste-treatment processes

It is clear that the first two steps imply avoidance of the generation of waste material and the two last steps mean loss of material resources. In practice, the waste-treatment processes usually include more than one of the steps. In reuse of packaging, a small percentage cannot be used and instead goes to material recovery or incineration. In the same way, re-refining of lubricants incinerates some 30–50 per cent of the oil while recovering the heat in the process. Incineration means often that around 30 per cent of the weight is deposited as ash in landfills or as filling material in construction.

There are cases of severe down-cycling that, at the worst, may be considered hidden landfills, as when high-grade materials, such as rubber or plastic, are used as filler in roads and construction. Ideally, recycled material should replace new material of the same kind or material of the same value or higher. In some cases, it may be difficult to compare one treatment process with another in this respect. The market value of the raw material replaced may be one way to differentiate the processes with regard to down-cycling. Studies on regional planning have shown that strategies favouring material recycling without down-cycling give lower total GWP than strategies favouring gasification (which in some respects is considered down-cycling into a fuel), which in turn gives lower GWP than incineration strategies [1].

In prioritising waste-management strategies, it is important to consider how the long-term development will proceed. Society and legislation aim to reduce the level of toxicity and other negative effects of waste and landfills; therefore, the cost of treating hazardous waste and maintaining landfills will rise faster in the long run. This makes it economically profitable to apply the same prioritisation in long-term business as in long-term political strategy. Thus, the best strategy for long-term economic and ecologically sustainable development is to redesign the processes, avoid the use of toxic materials, and avoid wasting materials and resources.

There is, however, a problem of differentiating between the steps in the approach, as in recycling a material that is a fuel or down-cycling a material into a high-value fuel.

There are alternative views on how to solve the problem [6, 7, 8][1]. Instead of avoiding the problem of mixing material recycling and energy recovery, this chapter discusses the few exceptions to the five-step approach.

When is the five-step approach not applicable?

Hazardous waste

For hazardous waste, all five steps are not always possible. Avoidance of toxic materials is always the best choice. Moreover, efficient long-term use of materials is preferable if it can be done safely and without adding more toxic material. However, some materials are unwanted in the eco-cycle. Therefore, there are usually strict regulations for the reuse, recycling and incineration of hazardous materials. Many organic materials can be incinerated and in this way lose their toxicity, while other materials cannot be totally incinerated and may have to be deposited. Laws usually require that the substances be safely deposited or destroyed and phased out. For example, mercury was phased out in 2003, carcinogens and compounds affecting genes and reproduction were phased out in 2007, and lead and cadmium will be phased out in 2010, according to Swedish law harmonised with EU regulations [5].

Down-cycling and heavy processing

Down-cycling has already been mentioned as a thing to avoid. In some extreme cases, down-cycling may be severe and mean that it would be more eco-efficient to incinerate the material with energy recovery than to recycle material with high environmental cost and low value. This may be the case when plastic is reused mainly as filler in building materials or recycling is done by an energy-intensive conversion process. In these cases, there is usually a better choice of material recovery. Another common example is when the material is a fuel or is down-cycled to a fuel. In this case, it may be hard to distinguish between material recycling and energy recovery, but then it is important to try to use the fuel in the highest energy grade possible; otherwise, the fuel is not used efficiently.

---

1   There are guidelines with other approaches than the five steps, most of which are directed to consumers. Some environmental organisations specify the first steps further and include use of recycled material in the waste strategy. The 'reduce' step is often divided into avoiding and prolonging life of the products [6]. Green Choices in the UK specifies the first step as 'refuse' unnecessary products, 'reduce' the amount of products required, 'refill' packages, 'repair' broken products and finally 'reuse' products internally or give them to someone else [7]. Sometimes it may be hard to give consistent guidance, since there are exceptions to the five-step stair. Even in recent thematic strategy on the prevention and recycling of waste (TSPR) and the revision of the Waste Framework Directive (WFD), the EU Commission fails to give a clear five-step approach and confuses material recycling with energy recovery [8].

Heavy material and long-distance transport

In applying different kinds of waste strategies, transport and handling emissions have to be considered. The proximity principle, that is, use material locally, should always be applied if possible. The handling and transport cost, as well as the environmental impact, must be considered in alternative treatment processes. In the few cases that material recovery requires great resources (energy or supplementary material) and long-distance transport of heavy material, incineration may produce less environmental impact than material recycling (Figure 2). However, in general, from an environmental point of view, there are few cases where the five-step prioritisation is not applicable due to transport. In most investigated cases, transportation does not have a significant environmental impact, especially when considering material of fossil-fuel origin.

Box 3. Reconditioning of sand and the transport impact

In 2002, sand reconditioning for foundries was investigated on behalf of the Volvo foundry in Sweden. The analysis showed that for the suggested location of the reconditioning plant, around 90 km from the foundry, the impact of transporting the sand would be worse than the impact of using it for landfill and construction material locally. In 2006, all of that material was used as construction material (down-cycling) locally.

Economic obstacles

From a short-term point of view, both incineration and landfill may often be cheaper than reuse and recycling. It has, for example, been cheaper to incinerate waste oil than to reprocess it into new lubricants. However, since the amount for reprocessing has been too low, legislation now requires reprocessing of oil even if it costs more. If secondary costs such as public relations and late implementation of legal requirements are considered, the five-step priority list usually turns out also to be a good economic rule of thumb.

Important application of the five-step approach

Do not incinerate fossil-based material

Many professionals trained after the early oil crisis of the 1970s claim that combustion of materials with energy recovery is the best option for fossil-based material. This was a valid claim when the aim was to remedy the shortage of fuel. Today, when the most important aim is to reduce the emission of carbon dioxide, incineration, with or without energy recovery, is not the best option. Recycling of plastic requires only one-seventh of the energy needed to make plastic from virgin oil, and production through recycling results in one-third of the carbon dioxide emission of production from virgin oil [9]. Furthermore, incineration generates fossil carbon dioxide in the process of burning, and any heat recovered in that process could have been generated by a non-fossil source. Similar figures are available for other fossil-based material. In addition, there are also usually economic advantages in avoiding incineration. For example, in Scandinavia, the cost of incinerating waste is 30–80 euros per tonne, while recycling usually generates revenue. Clean plastic residues from manufacturing plants may be worth up to 500 euros per tonne.

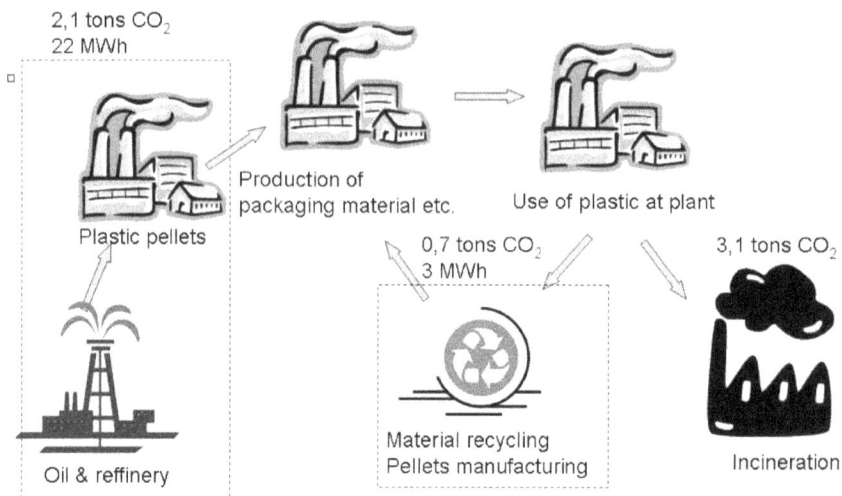

Figure 2. Comparing incineration of plastic waste and production of new plastic with recycling of plastic material (Hanarp and Kurdve, Volvo Technology, 2007)

However, a common argument against recycling and reusing materials and products is that the transport of the materials generates worse environmental effects than local landfilling or incineration. This has been proven true for only a very few exceptions. On the contrary, for most materials, transport produces much less impact than the process of making new materials from virgin resources. This is especially true for fossil-based materials such as lubricants, plastics and chemicals.

Box 4. Re-refining of waste lubricants at Volvo

A comparison of re-refining and incineration with energy recovery) of waste lubricant oil at Volvo plants in 2003 showed that, from an environmental point of view, re-refining is several times better than incineration. Transport within Europe did not produce a significant impact. The positive resulting impact came from avoiding production from virgin oil and transport. The total cost and revenue of sending wasted oil to oil-recycling was the same as for incineration in the short term; in the long term, it will be profitable.

Avoidance of wasting material

Chemical processes and surface treatment often generate waste solvents with significant environmental impact. Often these can be used as a high-grade fuel and disposed of at a low cost, and therefore their significance may be forgotten. Instead of incinerating solvent waste, it can usually be recycled into new products at equally low cost, and in most cases when a process change is possible, it may be reused internally on-site. The process redesign may need substantial effort. One way to reach a feasible economic and environmental solution may be to contract the supplier of paint and chemicals to take part in the design and quality guarantee of the cleaning and regeneration process. This has often been successful in 'chemical leasing' or 'CMS' partnerships where chemicals may be regenerated on- or off-site by the supplier, and the user pays for the service rather than for the product [10, 11].

Box 5. Internal reuse of solvent at a paint shop

At a paint shop, two alternative ways to dispose of cleaner solvents were considered: local incineration with energy recovery or distant, off-site recycling as solvent. The cost was equal, but avoiding incineration was found to have the better environmental impact. In looking deeper into the process, however, it was found that a less toxic solvent could be used and could be recovered and reused on-site. The effort in redesign of the process was substantial, but it is expected to give a bonus in a better working environment and less harmful environmental impact as well as savings in the long term.

Another important way to reduce the amount of waste is by the selection of packaging material. With increased demands of quality and farther transport of supply material, there is usually a constant increase of packaging material brought on-site. The increase means not only costly disposal but also internal logistic problems as well as more complex manufacturing operations. It is important to show these costs and problems to the purchasing and packaging departments. With a proper understanding of what the input of material costs, there may be significant gains of time, cost and productivity if packaging standardisation and common packaging by various suppliers can be enforced. The environmental gain can be seen as a bonus in the work towards higher productivity.

Box 6. Packaging substitution

Several assembly plants have had an increasing amount of cardboard and plastic, disposable packaging material. This makes more work for operators and may contribute to quality problems. Therefore, reusable plastic boxes have been introduced by several of the suppliers. By this extra logistic effort in the recycling system, there have been gains in reducing waste cost and transport, and increasing safety and quality at plant workstations and quality at the supplier.

Monitoring waste management

The five-step approach can be used as a general rule, but there is also a need to set up specific goals for how fast one should go up the steps for each type of waste. To achieve this, it is essential to have a good knowledge of what fractions of waste the operations generate and how these fractions can be treated. Basic goals for waste management could be set up, using the environmental aspect list from the environmental management system (according to ISO 14001 or similar regulations).

Box 7. Goals for waste management

Relevant goals for manufacturing industries

- reduction of total waste (TW)

- reduction of hazardous waste (HW)

- reduction of amount of non-product material use (TW/total material use)

- reduction of waste to landfill (W_{lf})

- increase of material recycling (W_{mr})

- 100% metal recycling ($^{metal}W_{mr}$)

- reduction of fossil-based material for incineration ($^{fossil}W_{er+inc}$).

These goals require different types of monitoring and different key performance indicators (KPIs). In many countries, legislation requires adequate monitoring and reporting on waste and hazardous waste (HW). In KPIs, the figure of total amount should be indexed per produced unit, per production hour, or by a similar standard. One should note that the definition and classification of HW change frequently in the legislation of different countries, and therefore they differs between regions. Thus, it is difficult to use HW/unit as a measure of overall environmental performance [12]. Another way to express the reduction of total waste (TW) is to measure it as a percentage of the total material used, that is, either tonne waste/tonne total material use (TW/TM) or tonne waste/tonne produced material (TW/PM), to monitor reduction of non-product material use. For these KPIs, the following formula can be applied: TM = PM + TW.

EU legislation requires reports on the treatment of the waste generated. This means that plants usually record how much of each type of waste is treated by incineration or landfill. EU legislation identifies some 15 types of final disposal (D1–D15) of waste and 13 types of recycling or energy recovery (R1–R13) [13]. These can, however, be grouped in 4 + 4 classes of waste: landfill, treatment ending in landfill, incineration or similar, energy recovery and material recycling, for each of HW and non-HW. When this is accomplished, the KPIs for each class can be determined. As with TW and HW, these should be indexed per produced unit, per production hours, or per tonne of products. It is also advisable to follow the cost equivalents of these KPIs as well as the volume measures.

Follow-up on goals for specific waste fractions requires more extensive data gathering than is required by legislation. With proper waste data management, however, usually all fractions are followed separately, and it is then easy to determine the origin of the material that is wasted. It may be difficult to define materials of fossil origin, but it might be a good idea to put all chemicals, oils and plastics into this group. The KPIs then might be a 'sum of volumes of all metal waste fractions per produced unit' and 'sum of all fossil-based waste fractions per produced unit'. Similarly, the indexes may also be based on production hours. It is also always advisable to calculate the cost equivalent KPI (Table 1).

Table 1. Suggested KPIs for waste management in manufacturing industries

	Landfill and similar	Treatment ending in incineration or landfill	Treatment with energy recovery	Material recovery	Produced material
k:	D1–D7, D12, D15	D8–D11, D13–D14	R1	R2–R13	–
Non-HW	ΣnonHW_k	ΣnonHW_k	ΣnonHW_k	ΣnonHW_k	–
HW	ΣHW_k	ΣHW_k	ΣHW_k	ΣHW_k	–
Total waste	ΣTW_k	ΣTW_k	ΣTW_k	ΣTW_k	ΣPM
Cost non-HW	$\Sigma \text{nonHW}_k (€)$	$\Sigma \text{nonHW}_k (€)$	$\Sigma \text{nonHW}_k (€)$	$\Sigma \text{nonHW}_k (€)$	–
Cost HW	$\Sigma \text{HW}_k (€)$	$\Sigma \text{HW}_k (€)$	$\Sigma \text{HW}_k (€)$	$\Sigma \text{HW}_k (€)$	–
Total cost	$\Sigma \text{TW}_k (€)$	$\Sigma \text{TW}_k (€)$	$\Sigma \text{TW}_k (€)$	$\Sigma \text{TW}_k (€)$	$\Sigma \text{PM}(€)$

10 tips/lessons

1. Try to understand the differences between the steps in the five-step approach and apply them to your residual material.

2. Set goals based on your industrial operations and follow them up with relevant KPIs.

3. Focus on large-volume fractions, high-total-cost fractions, hazardous fractions and landfill fractions.

4. Benchmark KPIs on volumes and costs with other plants with similar operations; discuss generation of waste and treatment alternatives.

5. Do not incinerate plastic and oil; sell them for a profit.

6. Take a look at your packaging material. Can you reduce it? Can you standardise it into one type of material for recycling?

7. Involve purchasers and packaging engineers in reducing waste-management costs.

8. When installing new processes, make sure that processed fluids and chemicals are recycled internally or externally, possibly in cooperation with the supplier.

9. Follow up on economic KPIs to ensure that future profitable investment comes through the decision-making process smoothly and successfully.

10. Even when everyone seems to work against you, do not give up.

References

[1]   Ljunggren, M. *A Systems Engineering Approach to National Waste Management*. PhD dissertation, Department of Energy Conversion, Chalmers University of Technology, 2000.

[2]   von Weizsäcker, E., Lovins, A.B. and Lovins, L.H. *Factor Four: Doubling Wealth – Halving the Resource Use* (London: Earthscan, 1997).

[3]   Volvo Trucks Corporation. 'Waste Management Guidelines for Dealers', Volvo Truck Corporation, Environmental Affairs reg.nr.20640/05-006 2005-08-18, Issue 1, 2005.

[4]   EPA. 'Quality Report for Statistics on Generation and Recovery and Disposal of Waste in Sweden in 2004 According to EU Regulation on Waste Statistics 2150/2002', Report 5594 June 2006, Swedish Environmental Protection Agency.

[5]   EPA. *A Strategy for Sustainable Waste Management – Sweden's Waste Plan* (Stockholm: Swedish Environmental Protection Agency, 2006).

[6]   Green Guardian. Waste management guide on the Internet, funded by Solid Waste Management Coordinating Board (SWMCB) and the Minnesota Pollution Control Agency, 2006 (www.greenguardian.com/business/rwmg_steps.asp).

[7]   Green Choices. Recycling guidelines by Green Choices, a non-profit-making limited company, 2006 (www.greenchoices.org/recycling.html).

[8]   RREUSE. 'Remarks from RREUSE on the Commission's Communication on the TSPR and the Revision of the WFD' – version II, 2006 (http://rreuse.org/33/fileadmin/ documents/00142_TSPR_and_WFD_Feb_06.pdf).

[9]   Nordin Håkan Miljökompassen A, Åi-Rapport 2002:1, 'Miljöfördelar Med Återvunnet Material Som Råvara', Återvinningsindustrierna, 2002.

[10]   Windsperger, A. 'The Variants of Chemical Leasing' in *Chemical Leasing: An Intelligent and Integrated Business Model with a View to Sustainable Development in Materials Management* (New York: Springer Verlag, 2004).

[11]   Mont, O. *Product-Service Systems: Panacea or Myth?* Doctoral dissertation, International Institute of Industrial Environmental Economics, Lund, Sweden, 2004.

[12]   Kurdve, M. and Mont, O. 'Chemical Management Services: Safeguarding Environmental Outcomes', Presentation at EMAN Conference, 'Environmental Management Accounting and Cleaner Production', Graz, Austria, 2006.

[13]   Directive 2006/12/EC of the European Parliament and of the Council of 5 April 2006 on waste (text with EEA relevance). Official Journal L 114 , 27/04/2006 P. 0009– 0021 (http:// eur-lex.europa.eu/LexUriServ/LexUriServ.do?uri=CELEX:32006L0012:EN:HTML).

12

Towards sustainable waste-management practices and education for sustainable development in Nigeria

Adewole Taiwo

Adewole Taiwo Associates, Nigeria

This chapter discusses a key dimension of 'WASTEnomics' – that of addressing sustainable waste-management practices and education.

Introduction

This chapter reviews the waste-management practices and the issue of sustainable development in Nigeria. The private sector, highway managers, local government and the Lagos State Waste Management Authority are mainly responsible for the collection and disposal of all types of waste generated in Lagos State. In terms of solid waste, only three dumps (erroneously referred to as 'landfills') exist in Lagos State, while all the closed dumps are still being used illegally, among several other illegal dumps that adorn the landscape. It was discovered that most of the industries, if not all, in the Isolo Industrial Environment of Lagos State have no pollution-abatement programme for their effluent, and that Lagos Lagoon alone is estimated to absorb 10,000 m³ of industrial effluent daily.

The waste-disposal habits of the people, corruption, work attitudes, and inadequate plants and equipment militate against effective waste management to attain sustainable development in Nigeria as a whole. Data generated by the study show that the method adopted by these agencies is ineffective and falls short of international standards in waste-management practices and sustainable development.

The findings of this chapter will be useful to researchers, government stakeholders and professionals working in the area of waste, waste to energy, metal recycling and climate change.

Waste management and sustainable development

The end of the 1980s saw a radical reappraisal of our concerns over resource availability and use, the environmental consequences of resource exploitation, and the relationship between the environment, poverty and economic change. This reappraisal has given rise to a new approach to environmental and development issues – an approach that seeks to reconcile human needs and the capacity of the environment to cope with the consequences of economic system. This approach is called sustainable development.

Sustainable development is an implied development without destruction; it is the judicious use of non-renewable resources for the present and future generations. Such non-renewable resources must be used at a judicious rate, to ensure that the natural wealth that they represent is converted into long-term wealth as they are used.

In Nigeria, we succinctly call it 'sustainable development without jeopardising future development', meaning that in our efforts to survey and exploit natural resources for our present purposes, there is a paradox evident in the need to ensure economic development while protecting the environment. It is important to note that there must be a balance between levels of development and the stock of natural resources; that is, development must be at a level that can be sustained without prejudice to the natural environment or to future generations.

Therefore, if there is to be sustainable development in waste management in Nigeria, land (for landfill), human resources, plant and equipment, and other factors, including capital, must be readily available. We need to ensure the future of the next generation by clearing our environment of all types of waste, taking into consideration both the physical and population development of the state. Therefore, waste management must mean the collection, storage, treatment and disposal of wastes in such a way as to render it harmless to human and animal life, the ecological balance, and the environment generally.

Waste

The Nigerian Federal Environmental Protection Act (1988) does not define 'waste'; however, waste, as the term implies, is any solid, liquid or gaseous substance or material that, being scrap, refuse or a reject, is disposed of or must be disposed of as unwanted.

One of the few statutes in Nigeria that attempts to define waste is the Lagos State Environmental Edict 1985, Section 32. Waste is defined as follows and includes:

1. waste of any description;

2. any substance that constitutes scrap materials or an effluent or other unwanted surplus substances arising from the application of any process.

The UK Environmental Protection Act 1990, re-enacting an earlier UK statute, took this statutory definition a step further in section 75(2), defining waste in the following terms:

Waste includes:

(a) any substance which constitutes a scrap material or an effluent or other unwanted surplus substance arising from the application of any process; and

(b) any substance or article which requires to be disposed of as being broken, worn out, contaminated or otherwise spoiled.

One thing to notice is that none of the above definitions of waste give 'value' to the elements considered. There is no suggestion that the items, which constitute waste do not have value or are intrinsically useless. The word 'unwanted', which appears in the definition, although it introduces its own problem, does not necessarily import a value element, for a substance or material may be unwanted notwithstanding that it has some value.

The issues of waste collection and disposal

Domestic waste management, collection and disposal have always been a universal problem. According to studies, it was noted that for years the major problem in Israel (especially in Ramat Hovar) was the accumulation of tens of thousands of tonnes of organic wastes. In the USA, too, until the 1970s, federal agencies had little authority to regulate hazardous waste, and solid disposal often took place in an unsafe manner at landfills or in banked lakes, some wastes simply being dumped on the ground or in surface waters.

Refuse and domestic waste are not a strange sight to Lagosians, whose streets are littered with tons of garbage, including animal and human carcasses. At present, private-sector waste-disposal operators diligently collect refuse bags from homes, load them into waiting trucks, and cart them away for final disposal. 'That is good,' the residents say. However, they worry that a lot of littering goes on, and the streets and avenues may be ignored and not cleared.

Studies have revealed that households account for about half, by weight, of the solid wastes generated in Third World cities, such as Lagos. It has also been noted that domestic waste-disposal management has received considerable attention not only in Lagos State but in Nigeria generally. Despite this laudable attention, the problems of collection, disposal processing, treatment, recycling and utilisation have defied solution.

The major problem caused by waste in the environment is pollution by various types of solid wastes, including paper, textiles, plastic, metals, glass, bone, wood, vegetable matter and food remnants of multiple consistencies.

It needs to be pointed out that the generation of waste materials is a problem that is not peculiar to Lagos alone. This problem is prevalent the world over, as noted earlier. This problem is not peculiar to the Third World alone but also affects the industrialised countries, where the pollutant effects of domestic and industrial wastes have caused considerable concern to environmental scientists. The problems in Nigeria arise from solid waste essentially. There are wastes from discarded materials generated by domestic and community activities or by industrial, commercial and agricultural operations.

Major classes of solid wastes

Municipal solid wastes generally can be classified in terms of three major sources: residential, commercial, and industrial. Sometimes, institutional sources are separated from commercial sources and, thus a fourth source is referred to as institutional. In the traditional scheme of classification, residential (domestic) solid waste consists of household garbage and rubbish, or refuse. The garbage fraction is mostly in the form of wastes derived from the preparation and consumption of food (e.g. meat and vegetable scraps). In the traditional scheme, all wastes that are not classified as 'garbage' are classified as 'rubbish'. The major constituents of rubbish include glass, metal, and plastic wastes; yard and garden debris; and wastepaper and paper waste.

It is against this backdrop that this chapter intends to review the impediments to effective and efficient waste management for sustainable development in Lagos State.

The major effects of waste on the quality of life

Environmental effects: The major environmental effects include air pollution, by odour, smoke, noise, dust, etc. Waste pollution also comes from disposal sites via flooding and overflowing because of blocked drains and land degradation.

Health effects: Health is affected by flies, which carry germs on their bodies and legs and also excrete them, and by mosquitoes that breed in stagnant water in blocked drains or in discarded cans, tyres, etc., that collect rainwater. Rats spread typhus, salmonella, leptospirosis and other diseases; they also cause injuries by biting and they spoil millions of tonnes of food. Refuse workers also face the hazards of parasite infection and cuts infected by skin contact with refuse, as well as, on the disposal sites, injuries from glass, razor blades and syringes, and tissue damage or infection through respiration, ingestion or skin contact.

Perceived causes of the intractable waste problem

There are many perceived causes of the intractable waste problem in Lagos State, among which are the following:

1. the waste-disposal habits of the people;

2. attitude to work;

3. inadequate vehicles, plant, equipment and tools necessary for waste management;

4. corruption;

5. overlap of function of the enforcement agencies;

6. population effect on waste management.

1. The waste-disposal habits of the people

Ignorance coupled with poverty may be adduced to explain the habits of most people in Lagos State, especially in the densely populated areas of the state. When we see a man defecating in broad daylight on the side of the highway, or a woman, with her dress pulled up, urinating on the sidewalk or gutter in full view of the public, or people throwing waste on the street from their cars, we can begin to grasp the dirty habits of the people. Nigerians are quite used to dirt. Evidence of this can be seen every day in the indiscriminate discharge of garbage into drains and at times on the highways.

A survey carried out on Lagos Waste Management Authority (LAWMA), which is charged with the responsibility of providing facilities for refuse collection, found that most of the streets in Lagos State do not have refuse collection bins – hence the irresponsible dumping of waste on the streets. Moreover, where public refuse bins are found, they are not regularly emptied.

Another survey found that the volume of municipal waste awaiting disposal is influenced by nearness to disposal sites, accessibility of transportation facilities, street layout, type of waste-disposal method and individual attitude. The individual attitude to waste disposal in Lagos State leaves a lot to be desired.

A situation in which a landfill that has been closed to the public is still being used as a dump calls for investigation. The fact that waste is dumped on the road median, inside gutter and roadside does not augur well for effective waste management.

Although communal dumps indiscriminately located in public places have been outlawed, illegal refuse collection points are commonly created by residents, presenting a health hazard and loss of environment aesthetics.

2. Attitude to work

In Nigeria, employee productivity is low due to certain factors, including a sociological factor, which is seen in the manifest lack of a sense of belonging to an organisation, and the tendency by employees to perceive a job as another's business. This poor attitude to work has harmed the waste-management efforts of the state government – the poor attitude to work, poor coordination and inadequate communication among workers and the institution saddled with solid-waste-management responsibilities, and bureaucratic impediments and administrative hitches have resulted in chaos, confusion and ineffectiveness in the delivery of many urban public services.

3. Inadequate vehicles, plant, equipment and tools necessary for waste management

Waste deposited at designated points of collection has to be transported either to the transfer loading station where sorting is done, or to the incinerator facility, sanitary landfill or other final disposal point. For an effective and efficient collection system, there must be enough well-maintained equipment such as tippers, pay loaders, bulldozers, road sweepers and compactors.

Problems such as inadequate number of vehicles, lack of spare parts, dearth of funds, poor technical know-how, poor maintenance practices, insufficient funding and lack of motivation have bedevilled the agency responsible for the disposal and collection of waste. The heaps and stretches of refuse that adorn the roads, pollute the environment, and disfigure the landscape are nothing but the result of inefficient waste-collection and management methods.

4. Corruption

Corruption is a canker that has eaten deep into every fabric of Nigerian society. This we may not deny except to our collective peril. The collapse of most LAWMA infrastructure in the state is alleged to stem from this menace of corruption. It has been sometimes reported that market women have had to bribe LAWMA operatives to remove waste from the market place. Cart pushers and scavengers have also been forced to bribe officials before they could dispose of their waste at the designated points – this has led to illegal dumps springing up and creating bottlenecks in the already chaotic situation of waste management.

5. Overlap of function of the enforcement agencies

Achieving sustainable development is inextricably linked to the establishment and enforcement of regulations, legislation and control criteria in environmental management and pollution control. But an overlap in the agencies responsible for effective enforcement of the various laws has created problems for effective waste management.

That LAWMA is saddled with the responsibilities of waste collection and disposal and must grapple with local government authorities, the Lagos State Environmental Protection Agency, the police and other enforcement agencies in the state does not augur well for effective enforcement and sustainable waste management. To have effective waste management and sustainable development in terms of waste collection and disposal effort in Lagos State, the enforcement mechanism should preferably be vested in only one organisation. Where many agencies need to be involved, their roles must be clear-cut and clearly spelt out.

The enforcement of environmental laws in Nigeria generally has been problematic. The management and regulation of the environmental laws have been beset by a host of problems, and have met with very limited success. These problems that hinder the enforcement of sanctions on violators of the environment are political, social and economic.

It is therefore clear that any effort towards a sustainable legal framework for successful enforcement, as well as avoidance of overlap of environmental laws, must come to terms with these issues, as a positive step towards the protection of the environment through effective waste management.

6. Population effect on waste management

Population growth and higher standards of living have always affected waste generation, collection and invariably disposal. The population of Lagos state rose from 1,443,569 in 1963 to 5,685,981 in 1991 and to 6,947,191 in December 1996. It is probable that the population of Lagos state has now reached 15 million. This has negatively affected both the environment and waste generation in the state.

Lagos State is the most densely populated state in Nigeria due to its commercial activities, and the quantity of waste generated in the state is in proportion to its population size – as population increases the waste generated also increases.

Like many other cities in the developing world, cities in Nigeria (especially Lagos) are faced with the twin problems of population increase and rapid expansion. These phenomena have, no doubt, caused increasing strain on the urban infrastructure. One area in which this strain has become obvious is waste management, as the existing system appears to be incapable of coping with the mountainous load of waste generated. Population growth goes hand in hand with increased pollution and environmental decay.

The way forward for waste management in Lagos State towards sustainable development

Some of the major problems militating against effective management and sustainable development of waste collection and disposal in Lagos state have been identified in the preceding section. These problems are hindering effective waste management and sustainable development. Therefore, there is an urgent need to develop an effective waste-management system for sustainable development.

Mitigating measures

Expanding recycling programmes can help reduce solid waste pollution, but the key to solving severe solid waste problems lies in reducing the amount of waste generated. Only the landfill system of waste disposal is being generally adopted in Lagos State, while other authorities employ several methods of waste disposal to ameliorate the problem of population effect on waste management. Some systems that could be adopted are as follows:

1. recycling;

2. bio-treatment;

3. incineration;

4. neutralisation;

5. secure sanitary landfill;

6. composting.

Moreover, international cooperation should be sought to learn how other countries have effectively managed their waste collection, handling and disposal. The state government should seize the opportunity to apply for international assistance in an effort to mitigate the looming disaster posed by population explosion in the state in terms of waste generation and disposal.

Remediation through education is also necessary. People should be educated on the need to reduce the amount of waste generated. The Lagos government should fund LAWMA to provide adequate collection bins in most areas of the metropolis and hinterland to forestall the bad habit of throwing waste anywhere and everywhere, creating the illegal dumps that adorn the major streets of the state.

A Sustainable Development Strategy Action Plan using a consensus-building approach should be formulated by government and other stakeholders, including the national government, the private sector, academics, environmental planners and experts, and non-governmental organisations. This action plan should last 20 years. It should be published and made available to the public, and fully implemented by all concerned.

As earlier mentioned, the waste-disposal habits of the people might change if the government stopped paying only lip-service to the serious issue of waste management. The availability and nearness of disposal sites would greatly inhibit the habit of dumping waste 'anywhere and everywhere'. Research and development into areas of better waste-handling methods might also go a long way to assist in improving the situation.

Nigeria has little or nothing to show as her achievements in the area of proper waste management. Heaps of garbage are commonplace along major roads, riverbanks, and ravines, and in excavated areas, particularly places excavated to obtain sand for road construction. In states where there is organised refuse collection, such as Lagos, the disposal of such wastes is usually on open dumps, located not far from living areas. Such dumps (called landfills) are not provided with environmental safeguards, and the leachates from them percolate freely into streams and the groundwater system.

Industrial effluents

With the exception of a few places, Nigeria cannot pride itself on having a functional sewage system. Industrial effluents of all types (both toxic and non-toxic) are discharged freely into surface and groundwater sources. Waste is allowed to pile up before it is ordered to be cleared with military dispatch and alacrity. This leaves room for corruption and does not allow effective waste management.

In developed countries, industries are obligated to discharge all their effluents into licensed, land disposal sites only, where such effluents are treated prior to reuse, recycling or discharge into streams or other approved places. There are no such controls in Nigeria, and where these sites exist, they are not regulated; most industries discharge raw, untreated and highly toxic liquid effluents into open gutters, drains, streams, lakes, estuaries and lagoons. For example, the effluent from the Aswani Textile Manufacturing Factory and other industries around the Isolo area in Lagos State is discharged on to the major road through the area. This helps to degrade that section of the road all year round and sometimes renders the road impassable, especially on the path of the effluent. Despite this, market women and men troop to this place every Tuesday to buy and sell – unaware of the great danger lurking there!

To most Nigerians, waste is merely a nuisance. They hardly give serious thought to the polluting effects of wastes and their deleterious effects on human health. Increase in urban population and 'blind', haphazard industrialisation has contributed a great deal to the generation of wastes in Nigeria. In the municipal areas of Lagos State, more solid wastes are produced than the generators can effectively cope with or manage. The situation of unmanageable wastes in the cities appears to worsen with perceived increase in the income of the inhabitants. The slums and the shanty neighbourhoods, as might be expected, receive little or no waste-disposal services.

Urgent solution for waste-disposal sites

We concentrate on two types of waste-disposal sites: landfills and open dumps. A *landfill* is differentiated from an open dump in that the landfill is an engineered design, consisting of a variety of systems for controlling the impact of land disposal on human health and safety and on the environment. An *open dump* is an uncontrolled system that has not been the subject of an engineering design (this is the type of waste disposal generally adopted in Lagos State and erroneously referred to as a 'landfill').

For our purposes, a waste-disposal site is generally defined to consist of that portion of the site wherein wastes are buried, as well as any surrounding property within the boundary of the site. The surrounding property may serve as a buffer, support landfill-related operation and facilities (e.g. maintenance) or unrelated activities (e.g. recycling depots), or contain access routes and roads.

Requirement for setting up landfills in Lagos State

Acceptable definitions of a modern landfill are based on the concept of isolating the wastes from the environment until the wastes are stabilised and rendered as innocuous as possible through biological, chemical, and physical processes. The main differences among definitions of a landfill involve the degree of isolation and the means of accomplishing it. Isolation includes preventing water from entering the landfill, as well as isolation of discharges directly from the landfill to the environment.

Three basic types of practices and requirements for a landfill are as follows:

1. consolidation of wastes into the work face; compaction of waste to conserve land resources; and design and operation of the landfill to control settlement, to optimise the chemical and biological processes (e.g. for landfill gas recovery), or both;

2. covering the waste with material on a daily basis to control the risk of hazards from exposed wastes;

3. control or prevention of adverse environmental effects from wastes disposed on land on soil, water, and air resources, and of the subsequent effect on public health and safety.

A landfill must meet the above three key conditions regardless of the stage of economic development of the country in which it is located. However, meeting the three conditions may be technologically and economically difficult or impractical in Lagos State. Therefore, the short-term or immediate goal should be to meet the conditions to the extent possible under existing circumstances, while the long-term goal should be to meet all three conditions eventually. This approach is recommended since the benefits associated with a modern sanitary landfill are realised only to the extent that a land disposal facility fully meets the three basic conditions. The most important condition is the prevention of negative effects on public health and the environment.

In conclusion, knowledge of the quantities and characteristics of the wastes to be landfilled is fundamental to the proper design and operation of a landfill. Among other things, these parameters influence or control many aspects of the landfill system over its lifetime, including the annual rate of filling, the required volumetric capacity of the fill, the production and characteristics of gases and leachates, and environmental effects.

The role of scavengers

The role of scavengers is very important in the planning, implementation, and operation of land disposal sites in Lagos State. First, the occurrence of scavenging between the point of waste generation and the disposal location influences the quantities of waste to be disposed of; therefore, this aspect of scavenging must be taken into account during the process of estimating waste quantities and characteristics in Lagos State.

Secondly, scavenging is widespread at existing land disposal sites in Lagos State (though this is prohibited in most developed countries) and is to be expected at new disposal sites unless policies and/or programmes are implemented to prevent the practice.

Scavengers are normally part of the socio-economic structure, and their displacement from a disposal site can have many direct and indirect consequences. While unsupervised and uncontrolled scavenging is detrimental to the health and safety of the scavenger, as well as to personnel operating the facility, the exclusion of scavengers from disposal sites is not necessary if their activities are managed and controlled.

Scavengers help in the recovery of valuable resources that would otherwise be lost as waste, and in the reduction of problematic materials, such as whole tyres and toxic chemicals, that enter landfills, and the subsequent adverse effect that they have upon landfill operation and performance. Scavengers also help to reduce the overall volume of materials in land disposal, and thereby help to conserve resources, such as land, and water and air quality.

The reason why Lagos State uses open dumps

In the case of developed nations, the degree of isolation of disposal sites considered necessary to protect the environment and human health and safety usually is much greater than would be technically and financially practical in many developing areas such as Lagos State. This high level of containment of a sanitary landfill, with its complex and expensive engineering system, as in a developed country, is simply beyond the means of Lagos State, and that is the main reason for its use of open dumps.

Urgent tips to attain best practice

- In terms of population, there is an urgent need for action plans and education in order to monitor and control waste disposal. There should also be room for international cooperation in achieving the action plans.

- Recycling must be expanded through the activities of scavengers and programmes such as Waste-to-Wealth.

- In landfill management and control, waste-to-energy programmes can be established, using the generation of methane gas.

- There is need to purchase up-to-date equipment, including vehicles, trucks, tippers, pay loaders, bulldozers and road sweepers, which must be backed up by a well-stocked maintenance store of spare parts. There is also an urgent need for well-trained staff.

- There is a need for a daily and regular refuse collection from both residential and industrial estates. There must be a disposal site in each street and avenue nearest to the sources of waste, and it must be accessible by everyone.

- The support of the private sector and non-governmental organisations is required, especially in the area of establishing maintenance workshops and enlightened programmes, which should include grass-roots participation and input.

- The federal and state governments should properly fund the various agencies responsible for effective and safe waste management. An agency in each of the various states should be responsible for the collection and disposal of all types of wastes.

- There is a need to build at least one domestic waste incineration plant for each local government council with a daily waste treatment capacity of at least 3,000 tonnes. These plants must be able to generate electricity by incineration so as to recover the energy of the waste.

- There is need for more effective, up-to-date domestic waste-transfer stations.

Section 4:
Towards Zero Waste

13

Achieving zero waste through waste prevention

Sandra Lebersorger

Institute of Waste Management, Department of Water, Atmosphere and
Environment, BOKU – University of Natural Resources and Applied Life Sciences,
Vienna, Austria

This chapter discusses a key dimension of 'WASTEnomics' – that of achieving zero waste by waste prevention in industry and by consumers. The clear benefit of waste prevention along the entire life cycle of a product, including reduction in resource consumption (energy, water and material) as well as emissions (to air, water and solids), has resulted in recent concepts such as cleaner production and zero emissions.

Introduction

Local communities are facing a growing waste burden that is becoming increasingly more difficult to manage. The problem here is not only the quantity of waste but also the quality, that is, the intrinsically hazardous nature of some types of waste, especially industrial waste. Industry today generally uses a wider range of materials and produces more complex products than in past decades. There has also been an overall increase in the quantity and variety of products and services and a continual creation of new products. All these trends, together with the fact that products are traded worldwide, have also increased the number and variety of players involved, and this more than ever makes waste prevention a challenging matter.

In view of the increasing waste quantities, waste prevention has been assigned the highest priority under European waste management law. In the past, various efforts were made at both national and international level to institutionalise waste minimisation and waste prevention by setting legal guidelines to attain effective waste prevention. Based on the Council Directive of 15 July 1975 on waste (75/442/EEC), the first Community Strategy for Waste Management (SEC(89) 934 Final 1989) established the hierarchical system of waste management, under which waste prevention and minimisation were given the highest priority, followed by recycling and disposal. This development was continued in the Community Strategy for Waste Management of 1996 (COM (96) 399), in the Thematic Strategy on the Prevention and Recycling of Waste (COM(2005) 666 Final), and in a Communication on Integrated Product Policy (COM(2003) 302 Final).

However, the initiatives which have been taken so far have not reduced the regular annual increase in total waste. Can this wide gap between the paramount significance of waste prevention in policy measures and the steadily increasing amounts of waste be explained by the absence of a comprehensive strategy, too little effort spent on the implementation of waste-prevention measures so far, conflicting interests, or simply the problem of measuring the short-term effects of the long-term waste-prevention process? Is waste prevention only a forward-looking philosophy?

Definition

The first difficulty when talking about 'waste prevention' is to ensure that everyone comprehends the meaning of this term in exactly the same way. Although the terms 'waste prevention' and 'minimisation' are commonly used, there are no strict definitions of them.

Figure 1 illustrates how waste prevention is defined in this chapter. The definition is based upon the OECD working definition at the Berlin meeting in 1996 [1]. Measures were classified into preventive measures – prevention, reduction at source, reuse of products – and waste-management measures. Waste minimisation includes the preventive measures as well as the waste-management measures, 'quality improvements' and 'recycling'. 'Waste prevention', as a generic term, means reduction in the amount of generated waste (quantitative prevention) as well as reduction in the hazardousness of waste (qualitative prevention) and includes both preventive measures and also 'quality-improvement' measures.

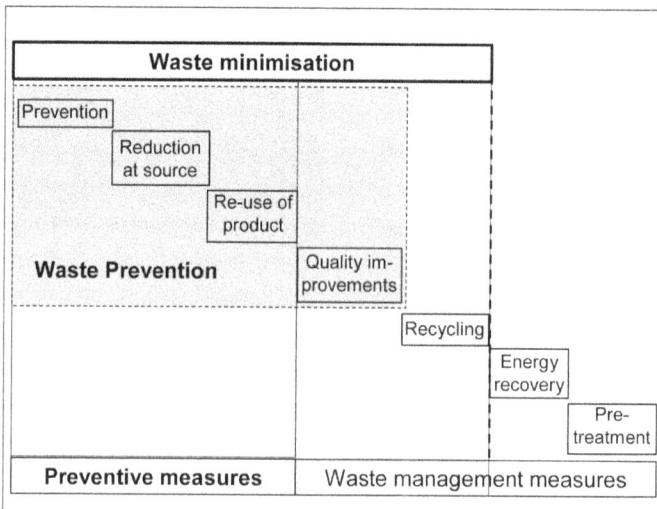

Figure 1. Definitions in waste prevention (OECD working definition, modified)

Instruments for initiating waste prevention

Many instruments are available to initiate or advance waste-prevention measures. These instruments can be divided into five groups (Table 1), referring to the basic way in which an agent chooses to influence the behaviour of another agent. By contrast, the term 'measure' is used to indicate the concrete realisation of an instrument (for example, providing a brochure with addresses of repair shops in a city to every household). Very often, a combination of different types of instruments is used or even necessary (for example, providing a subsidy for reusable diapers and informing the target group about this; announcing a tax and creating the legal basis for it).

Table 1. Typology of instruments (Kaufmann-Hayoz et al. [2])

Instruments	Examples
Command and control instruments	Emission standards, product standards and regulations, licensing, regulations
Economic instruments	Subsidies, incentive taxes, charges, deposit-refund systems, market creation, incentives
Collaborative agreements	Public–private agreements, certification and labels
Service and infrastructure instruments	Second-hand markets, sharing networks, repair shops
Communication and diffusion instruments	Brochures, direct personal contact, mass media, without a direct request or with a direct request

Command and control instruments are mandatory regulations which restrict the range of options for the target group. In order to enforce regulations, economic measures such as fines are used, often additionally. Command and control instruments can only be implemented by legislative authorities (national, federal or municipal authorities), but can be used to influence the behaviour of any target group. The implementation often requires great technical effort and economic costs. For example, as a consequence of ROHS directive (2002/95/EG), which limits the use of hazardous components such as mercury in electrical and electronic equipment, producers have to adapt their technology while legislative authorities have to incur economic costs in enforcement purposes.

Economic instruments can be implemented by any authority. They aim to direct behaviour of the target group – industry as well as consumers – towards environmentally sound behaviour by raising the cost of environmentally harmful behaviour, reducing the cost of environmentally sound behaviour, or creating markets for tradable permits. Economic instruments should be flexible and simple. Control and enforcement are essential for their successful implementation. Measures for waste prevention include, for example, a tax on beverage packaging, a subsidy for the purchase of a basic package of reusable diapers, or tax relief for industries which use low-emission technologies.

Collaborative agreements are legally binding or non-binding commitments by the private sector (industries, associations or individual companies) to the government. The target group is involved in the definition of the agreement, which can include the definition of a certain goal (for example, guaranteeing that 80 per cent of beverages are sold in refillable packaging) or of measures. Collaborative agreements are often used in sectors that are traditionally not regulated by law. The threat of more restrictive legal measures, if a collaborative agreement cannot achieve its goal, is often used to force industries to participate.

Service and infrastructure instruments are goal-directed transformations of services or infrastructure for the promotion of certain behaviours. They determine to a large extent what actions are or are not objectively possible for certain agents [2]. They can be applied by any agent and to any target group. The effort for their implementation depends on the scope of the measure. While larger projects require a political basis, measures on a small scale can be easily implemented. Examples in waste prevention are the reuse and marketing of waste products that are still useable or simply require repairing, via second-hand shops, repair shops or sharing networks. Such initiatives often also have a social purpose, as by employing people who have been out of work for a long time or offering low-priced goods to socially disadvantaged people. In Vienna, the R:U.S:Z (repair and service centre) collects and repairs second-hand washing machines, dishwashers, dryers, TV sets, hi-fi systems and video equipment, and resells them.

Communication and diffusion instruments can be applied by any agent and to any target group, in conjunction with other instruments or in the initial phase of the implementation of a waste-prevention strategy. Communication instruments aim to influence the internal conditions of the target group – its aims, knowledge and behaviour – and refer to the contents of a message. Communication instruments impart information or values, without a direct request (for example, presenting model behaviour by showing families who try to minimise their waste quantities; presenting ways to prevent waste) or with a direct request (for example, sending appeals not to buy non-returnable bottles or products with excessive packaging). Diffusion instruments refer to the way in which a message is distributed: via direct personal contact, person-to-person media or mass media.

Waste prevention in industry

Waste prevention in industry is usually closely connected to the prevention of liquid emissions into water and gaseous emissions into the atmosphere. Most concepts therefore consider not only waste, but also all kinds of emissions. Historically, the first step was end-of-pipe solutions aimed at reducing the amount and hazardousness of emissions after they had been created. The focus then switched to the consideration of environmental effects along the entire production cycle (cleaner production concepts) and finally to zero-emissions concepts that propose that material cycles from intake to emissions should be managed as a holistic system [3, 4].

There are a lot of opportunities to advance waste prevention in industry. Command and control instruments, such as emissions standards or restrictions on the use of pollutants, can be applied by legislative authorities. Special conditions can be imposed upon an industry in the context of the permission for a plant granted by an authority. Given that the primary financial purpose of industry is to be profitable, the potential reduction of cost can be an essential driver for the optimisation of production technologies and the minimisation of the quantity of waste. On the other hand, a significant barrier to waste prevention is the widespread view of industry that environmental regulations of industrial activity and waste reduction are costly for companies and have a negative impact on them [5, 6]. For example, the disposal of hazardous waste is often a significant expense factor.

Environmental management systems

Environmental management systems and certifications, such as EMAS (eco-management and audit scheme) and ISO 14001, provide a basis for the realisation of the environmental goals of a company. Key elements of EMAS are the implementation of an environmental management system, the objective and systematic evaluation of its performance, the communication of information to the public, and the involvement of the employees. Furthermore, the concept requires the use of the best available technologies and continual improvement of the company's environmental performance. The benefits of an environmental management system for a company are cost savings, better competitiveness, a positive corporate image, and advantages when dealing with authorities (Box 1).

Box 1. Effect of an environmental management system

A case study [7] investigated the effect of an environmental management system on waste quantities and costs in 31 automobile service shops with fewer than 100 employees. The automobile service sector generates, both qualitatively and quantitatively, significant volumes of waste. Roughly half of the waste generated consists of hazardous fluids, including waste oil, waste solvents and used brake fluids; in addition, there are mixed wastes, such as hazardous shop waste, which can be disposed of only by incineration, and residual wastes. Within eight years, the following measures had been implemented by the shops:

- recycling and reuse systems: for example, reusable cleaning tissues, reusable seat covers, reusable oil suction mats, refillable spray cans, parts- and paintbrush-washing machines with cleaning liquid recycling, and separate collection and recycling of windscreens and catalytic converters;

- technical innovations: central electric oil supply and waste oil suction, and use of water-based paints;

- information and training: work instructions, appointment of a waste representative, general environmental protection training, and documentation;

- cost-reduction measures: optimisation of disposal intervals and price negotiations with disposal companies.

Comparison of the waste quantities and costs before and after the implementation of the environmental management system showed that, on average, the total quantity of waste could be reduced by 12 per cent (Figure 2), and costs could be reduced by 28 per cent. There was an especially significant reduction of (cost-intensive) hazardous waste, clearly illustrating that the introduction of an environmental management system has economic and environmental benefits for companies.

Figure 2. Waste quantities and costs per car before and after implementation of measures

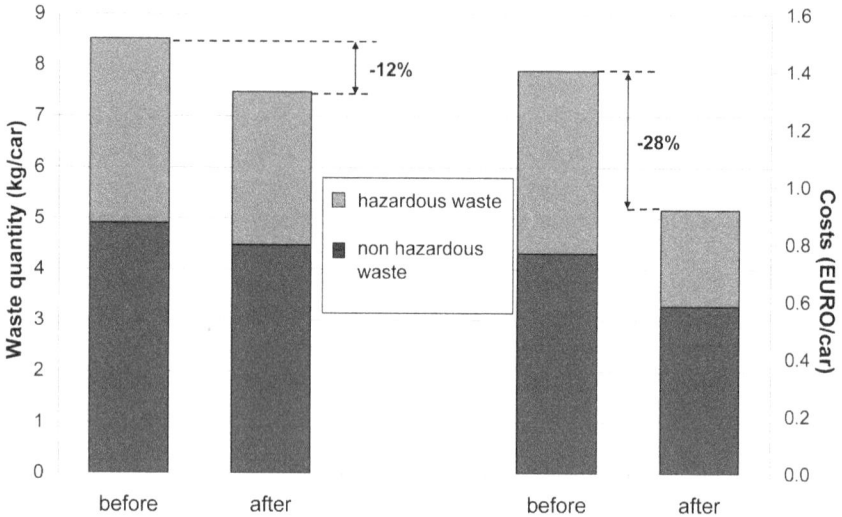

Cleaner production

The cleaner production concept is an integrated preventive strategy for the entire production cycle and considers environmental, economic and social aspects. The targets of cleaner production are to increase productivity, to promote better environmental performance, and to reduce the environmental impact of products. The strategies to reach these targets include more efficient use of raw materials, energy and water; reduction at source of waste and emissions; and the design of environmentally friendly but cost-effective products (eco-design). Analysis of material and energy flows in industry should lead to organisational and technical measures to optimise production processes. Examples of such measures are the documentation of consumed materials, the definition of benchmarks, the application of new technologies that produce less waste, the recycling of wastes, careful selection of raw materials and auxiliary materials, the extension of the utilisation phase of auxiliary materials, and improving automatic control.

The 'cradle to cradle' perspective [8] envisages that products or materials be designed in such a way that after their useful lives they either re-enter nature without leaving synthetic materials or toxins, or circulate as pure and valuable materials within closed-loop industrial cycles rather than being down-cycled into low-grade materials.

The cleaner production concept has been implemented in many national and international programmes. In Austria, the initiatives, Prepare and Oekoprofit, were started in 1992; in Germany, the initiative PIUS was launched in 1999. In an initiative by the United Nations Industrial Development Organisation (UNIDO), 43 national cleaner production centres and programmes have been established in Africa, Asia, South and Central America, and Europe. The initiative aims to build national capacities for cleaner production in developing countries and countries in transition, to foster dialogue between industry and government, and to enhance investment for transfer and development of environmentally sound technologies to developing countries (www.unido.org).

Cleaner production programmes can increase competitiveness, facilitate market access, and strengthen the productive capacity of industries by reducing costs and improving environmental and social performance at the same time, as China's pulp and paper industry illustrates [9]. Fifteen pulp and paper mills with 700–4,500 employees participated in this programme between 1995 and 1997. This industry causes high environmental pollution, and black liquor, which generates 80–90 per cent of the overall pollution load, is the main environmental problem. A number of different measures were implemented, such as the optimisation of technologies and processes (for example, improving black liquor extraction by higher vacuum, temperature change, and improved bleaching), modification or updating of equipment, on-site reuse, recycling and recovery, raw-material substitution, and training and incentives. Besides significant reduction of resources (water, coal, electricity and alkali) and emissions (waste water), the quantity of solid wastes from 11 mills was reduced by 1,000 tonnes per year. The total cost saving for all 15 participating mills was 52.1 million RMB/year (4.8 million euros).

Zero emissions

Unlike the traditional linear industrial models in which wastes are considered the norm, the zero-emissions concept envisages an integrated, holistic system in which everything has its use, and all industrial inputs are used in final products or converted into inputs for other industries. This requires the reorganisation of industries into clusters in which each industry's wastes or by-products are fully matched by the input requirements of another industry. Thus, the manufacturing line can be viewed as integrated technologies and series of production cycles and recycling systems [3].

The zero-emission concept does not assert that all emissions can reach precisely zero. It is an idealised goal, forcing a systems perspective and not only concentrating on materials, but also referring to management standards and symbolising a process of continual improvement. The primary focus is the intake of natural resources within acceptable limits and final emissions within acceptable limits [4].

In 1994, the Zero Emissions Research Initiative (ZERI) was introduced by the United Nations University. This international scientific programme is designed to investigate approaches and technological breakthroughs as a basis for the creation of a new type of industrial system. The current focus is upon renewable biomass as a substitution for fossil fuels, using biorefinery [3]. A multitude of developments are aiming for zero emissions in Asia, Europe and North America, as, for example, the utilisation of lignin from paper production as a substitute for plastic in the automotive and cosmetic industries. However, critics argue that a demonstration project for zero waste does not yet exist, since most of the projects focus upon solutions by engineering and natural sciences, but do not take into consideration the social, political and economic framework [4].

The zero-emissions concept also requires a shift in society towards consumers who preferentially purchase functions instead of materials [4]. Another phenomenon which has to be dealt with, is the 'rebound effect': although substantial improvements in the environmental efficiency of production processes have been achieved in recent years, the net effect of these technical innovations has been reduced by increased consumption in most countries and within most sectors [10]. Moreover, consumer behaviour in the utilisation phase has also to be taken into account. Life-cycle assessment (LCA) studies show, that for a number of products, the use phase is more important than the production phase [10], as, for example, in terms of energy consumption.

Waste prevention by consumers

Inducing consumers and private households to prevent waste is probably more difficult than to motivate industry. One basic difference can be found in the purpose of households. Unlike industry and business, which primarily seek to make profit, private households aim to satisfy the physical and non-physical needs of their members. Command and control instruments for waste prevention are far more difficult to apply to private households, since their enforcement and control are often hardly possible. Moreover, economic incentives are not sufficient to overcome all barriers, as the example of reusable diapers (Box 2) illustrates.

Box 2. Reusable versus one-use diapers

In Austria, several local waste authorities grant subsidies to encourage parents to buy reusable diapers. Usually, the purchase of the basic items for reusable diapers at a price of 250 euros or more is subsidised by 100 euros. Even without this incentive, the total cost of reusable diapers during infancy is far less than that for single-use diapers (Figure 3). Nonetheless, current participation rates are very low – the incentive is made use of for less than 5 per cent of newborn babies. Obviously, there are other barriers, such as the greater need for information about the handling of reusable diapers, the effort which is perceived to be greater, or the competition from the commercial promotion of single-use diapers, which appear much stronger than the prospective cost savings.

Figure 3. Costs incurred by households when using reusable or single-use diapers [11]

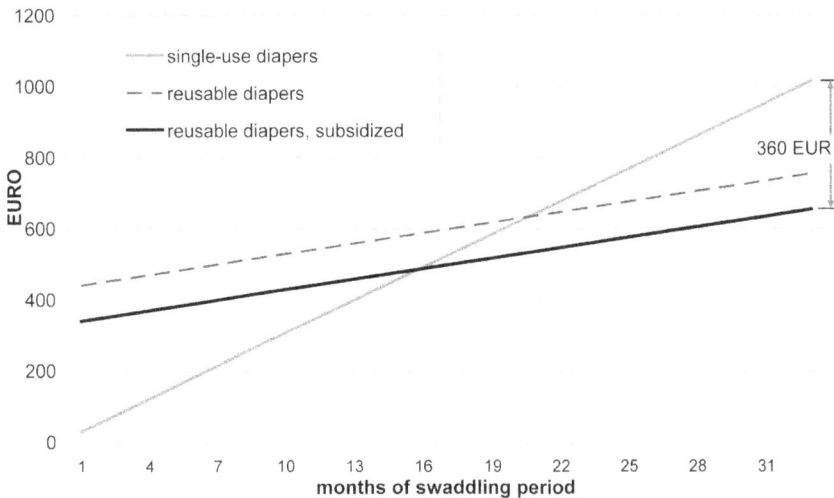

Previous measures for waste prevention usually used information and diffusion instruments and primarily addressed the prevention of packaging, the substitution of one-use packaging by reusable packaging, reuse and repair of products or the refusal of advertising material. A reduction of consumption which is often considered to be the only choice in view of sustainability [12], has not been picked out as a theme so far. On the one hand, reducing consumption is contradictory to our modern economic system and society – that is why politicians have not yet dared to address this thorny subject. On the other hand, reducing consumption will require a significant change of people's habits and thus face significant barriers. Habitual behaviour, which constitutes the main part of our everyday life, has an essential function for the individual as a protection against stimulus satiation. Initiating changes is a great challenge, particularly since most social conditions now support a 'wasteful' way of life – for example, the increase in the number of single households, general shortage of time, increase in complexity, transition to a more flexible society, and individualisation.

In Austria, composition analyses of household waste indicate that in recent years increasing amounts of unconsumed foodstuffs, often still in their original packaging, have been thrown out by private households (Figure 4). These wastes amount to up to 39 kg per capita a year or up to 12 per cent by weight of residual waste. In the UK, the amount of food waste is estimated to amount to 19 per cent by weight of household waste or 6.7 million tonnes per year (www.wrap.org.uk). Upon initial examination, there seems to be a 100 per cent theoretical prevention potential for wasted food, since it does not represent unintended by-products of consumption (compared to packaging or residues from food preparation) but rather products for which money has been spent but which have not been used for the intended purpose (eating or drinking).

However, the fact that the purchase of food is predominantly habitual behaviour and that the act of buying also involves a variety of physical and non-physical needs makes the prevention of food seem a much more complex task. Changing consumer behaviour to a more sustainable consumption lifestyle will therefore be the great challenge of the future.

Figure 4. Food recovered from residential waste (picture: ABF–BOKU)

Challenges and implications

Given the increasing waste quantities worldwide, waste prevention must be considered a priority solution. This chapter discussed different aspects of waste prevention in industry and by consumers. The key points can be summarised as follows:

- The relevance of waste prevention and an integrated product policy has been recognised by local, national and international legislative authorities.

- As a consequence, a legal basis has been established, and many initiatives to foster waste prevention have been started. But no appreciable reduction in total waste quantities can be noticed so far. The quantities of municipal solid waste in Europe, as elsewhere, are still increasing.

- A problem thereby is how to evaluate waste prevention. Usually, the mass or volume of waste is used as an indicator. Relevant baseline data are often not available.

- Utilising only end-of-pipe solutions is rather myopic. There are several other beneficial effects from waste prevention along the entire life cycle of a product, including reduction in resource consumption (energy, water and material), emissions (to air, water and solids) and social aspects. This has been taken into account in recent concepts such as cleaner production and zero emissions.

- Examples show that the application of cleaner production concepts or environmental protection measures not only generate significant reductions of emissions and resource consumption, but also reduce costs for industry. This is a great economic and strategic advantage in encouraging waste prevention.

- While significant reductions of production waste have been achieved in recent years, consumers are far more difficult to influence. Consumption is still increasing. Except for some simple initiatives (such as 'no, thanks' stickers for advertising material and prevention of excessive packaging), consumers can prove difficult to motivate to prevent waste, often not even by economic incentives, because of barriers such as habits, perceived effort or modern social conditions. Changing consumer behaviour to support more sustainability will be the great challenge of the near future.

- If waste prevention is to be effective in the long term, an integration of all agents along a product's life cycle is required as well as the integration of waste prevention into society and its values. Concepts such as zero waste mark an important first step in this direction.

References

[1]   EEA (European Environment Agency). *Case Studies on Waste Minimisation Practices in Europe*. Topic report 2/2002 (Copenhagen: EEA, 2002).

[2]   Kaufmann-Hayoz, R., Bättig, Ch., Bruppacher, S. et al. 'A Typology of Tools for Building Sustainability Strategies' in Kaufmann-Hayo, R. and Gutscher, H., eds. *Changing Things – Moving People* (Basel: Birkhäuser, 2001) pp.33–108.

[3]   Gravitis, J. 'Zero Techniques and Systems – ZETS Strength and Weakness', *Journal of Cleaner Production*, 2007, 15(13–14), 1190–7.

[4]   Kuehr, R. 'Towards a Sustainable Society: United Nations University's Zero Emissions Approach', *Journal of Cleaner Production*, 2007, 15(13–14), 1198–1204.

[5]   Chapple, W., Morrison, P.C.J. and Harris, R. 'Manufacturing and Corporate Environmental Responsibility: Cost Implications of Voluntary Waste Minimisation', 2005, *Structural Change and Economic Dynamics*, 16(3), 347–73.

[6]   Beaumont, N.J. and Tinch, R. 'Abatement Cost Curves: A Viable Management Tool for Enabling the Achievement of Win–Win Waste Reduction Strategies?', *Journal of Environmental Management*, 2004, 71(3), 207–15.

[7]   Starke, R. 'Optimierung der Abfallwirtschaft von KFZ-Reparaturbetrieben durch freiwillige Umweltschutzmaßnahmen' ('Optimising Waste Management of Automobile Service Shops by Means of Voluntary Environmental Protection Measures'). Dissertation, BOKU – University of Natural Resources and Applied Life Sciences, Vienna, 2002.

[8]   McDonough, W. and Braungart, M. *Cradle to Cradle: Remaking the Way We Make Things* (New York: North Point Press, 2002).

[9]   Xin, R. 'Cleaner Production in China's Pulp and Paper Industry', *Journal of Cleaner Production*, 1998, 6, 349–55.

[10]   Throne-Holst, H., Stø, E. and Strandbakken, P. 'The Role of Consumption and Consumers in Zero Emission Strategies', *Journal of Cleaner Production*, 2007, 15(13–14), 1328–36.

[11]   Salhofer, S., Obersteiner, G., Schneider, F. and Lebersorger, S. 'Potentials for the prevention of municipal solid waste', *Waste Management*, 2008, 28, 245–59.

[12]   Fricker, A. 'Waste Reduction in Focus', *Futures*, 2003, 35(5), 509–19.

Further References

Salhofer, S., Graggaber, M., Grassinger, D. et al. 'Potentiale und Maßnahmen zur Vermeidung kommunaler Abfälle am Beispiel Wiens' ('Potentials and Measures for the Prevention of Municipal Solid Waste, Exemplified for Vienna'), Beiträge zum Umweltschutz Heft, 2000, 67(1).

14

Where is waste heading? Predicting waste generation to ensure accurate capacity planning

Peter Beigl

Institute of Waste Management, Department of Water, Atmosphere and Environment, BOKU – University of Natural Resources and Applied Life Sciences, Vienna, Austria

This chapter discusses a key dimension of 'WASTEnomics' – that of understanding the socio-economic factors that influence the amount and composition of future waste streams. The use of improved waste generation enables more adequate dimensioning of the waste collection infrastructure and treatment facilities, thus helping to avoid expensive over- or under-capacity.

Introduction

An understanding of the socio-economic factors that influence the amounts and composition of municipal solid waste (MSW) allows improved forecasts as a basis for waste management planning. However, accurate forecasting of MSW quantities is a highly uncertain, if indispensable, element of waste management planning. Although capacity planning of waste-processing facilities and infrastructure requires an idea of the future demand, little is known about how to estimate the quantity of future MSW streams. This neglect of the first step in the planning process is likely to provide the wrong answers to questions such as 'Do we need another new, enlarged waste incineration plant, recycling facility or composting plant?' A growth forecast that is out by only 1 per cent can lead to a deviation of more than 10 per cent of the total waste generated over a planning period of 10 years. Under- or overestimation thus has significant consequences in terms of additional investment and operating costs.

This chapter describes the development of a forecasting method to enable more accurate estimation of the amounts and composition of MSW generated in the future in European cities over a period of at least 10 years. The work was carried out as part of a project, 'The Use of Life Cycle Assessment Tools for the Development of Integrated Waste Management Strategies for Cities and Regions with Rapidly Growing Economies' (LCA-IWM), initiated within the thematic programme, 'Energy, Environment and Sustainable Development' of the European Commission (EC)'s Fifth Framework Programme. The main aim of the first part of this project was to provide a reliable decision-support tool for long-term waste management planning in European cities (particularly its application in rapidly growing economies).

Extensive investigations led to the identification of a set of suitable, prosperity-related indicators of waste generation in a long-term perspective. Consideration of underlying socio-economic trends – expressed by variables such as the gross domestic product (GDP), household size and health indicators – could help to improve the understanding of these relationships and thus to improve forecast accuracy in the long term.

Learning about future MSW – what for?

The way towards zero waste and a circular flow economy is paved with many decisions in order to restructure the material flows and processes. Some of them are strategic decisions about capital-intensive investments, such as MSW incinerators, recycling plants or new waste collection infrastructures. The recycling activities adopted and changing legislation are at least two aspects that can increase the uncertainty in long-term planning. The long time lag between decision and start-up of a treatment plant complicates the planning process. The main investment decisions of private and public investors can be outlined in the following three questions.

Which technology to select?

A well-funded commitment to a treatment technology is based on knowledge about present and future quantities of waste streams that have to be treated. Especially in regions with rapidly growing economies, it is important to consider both the significant growth of overall MSW generation (including recycled and disposed fractions) and increasing recycling quotas. Overview of the expected developments for each waste material will help in assessing whether a technology is promising or not.

Which design capacity?

Anticipatory planning procedures make assumptions about the different developments of various influencing factors, such as economic and population growth under optimistic or pessimistic conditions. Up to now, the quantification of the relations between waste generation and potential influencing factors has hardly been investigated. Thus, the determination of the design capacity is often based only on estimations from historic data. If quantitative forecasts are used, simple trend extrapolations are carried out by using historic time series [1].

Which quantities from which suppliers?

Contractual obligations play an important role in long-term capacity planning. Either an excess or a shortfall of input allotments can affect the utilisation of a plant. Prospective assessment of future waste quantities to be processed can help to develop measures to overcome such problems in time.

Data sources

In order to focus on regional waste management systems in environments with different levels of economic growth, the study within the mentioned EC project (LCA-IWM) considered 91 cities in 44 European countries, including all capitals and cities with more than 500,000 inhabitants. Led by the Institute of Waste Management at BOKU – University of Natural Resources and Applied Life Sciences in Vienna, waste-related data, as well as demographic and socio-economic indicators at a city level, were collected between October 2002 and February 2003 in cooperation with local representatives. To enable the analysis of long-term changes in waste-generation patterns, available data were collected from 1970 to 2001. The collected information was examined in collaboration with six project partners from various European countries (see Acknowledgements).

Waste database

As well as the total MSW stream, our primary interest was the generation (not the collection!) of material-related fractions (paper/cardboard, organics, glass, plastics and composites, and metals), regardless of whether these were collected separately or as part of commingled waste. The availability and comparability of international waste statistics are key issues in studies such as this. With this in mind, the project focused on all waste streams from daily and routine activity of households and businesses, excluding sporadically collected wastes such as construction and demolition waste. Use of this definition (as defined by the European Environment Agency [2]) meant that it was possible to evaluate data with an average time-series length of 10 years on MSW quantities from 55 European cities in 10 old member states and five new ones, as well as two applicant countries. Data from other cities either were not available or did not pass the plausibility check with regard to comparability. Only 45 datasets from 31 cities were available for material-related waste generation.

Socio-economic indicators

The level of prosperity is generally regarded as the main factor affecting the amount of MSW generated per person. Numerous previous studies have tested the relationship between prosperity and related proxy variables. These have shown that some indicators (e.g., GDP and household consumption expenditure) correlate positively with the quantity of waste generated, while others, such as household size, are typically negatively correlated.

More than 30 demographic, economic and social indicators were selected for collection at a city and a national level. Of these, the indicators listed in Table 1 were available as time series. These data were obtained from many regional, national and international statistical sources (e.g., Eurostat, United Nations, Organisation for Economic Co-operation and Development (OECD)) through database investigations, literature and Internet reviews, and personal communication.

Table 1. Available socio-economic indicators

Available indicators at city and national level

- Population

- Population density

- Population age structure (0–14/15–59/≥60 years)

- Sectoral employment (agriculture/industry/services)

- Gross domestic product/regional product

- Infant mortality rate

- Life expectancy at birth

- Overnight stays (tourism)

- Average household size

- Unemployment rate

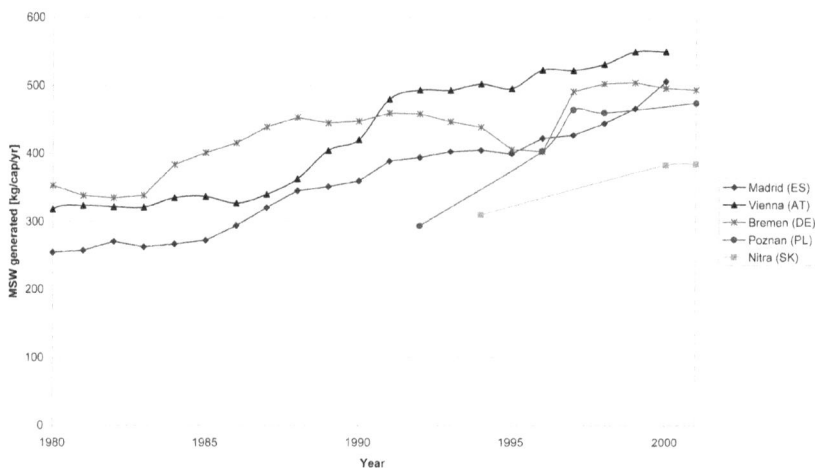

Figure 1. MSW generation in selected European cities, 1989–2001

Increasing amounts of waste

Apart from intermittent decreases, generation rates grew steadily throughout Europe between 1995 and 2001 (Figure 1). The data for the eastern European cities of Poznań and Nitra suggest that the gap between old and new member states is closing.

A comparison of all the 43 cities studied in old EU member states (EU15) and 12 cities in central and eastern European (CEE) shows clear differences between these, until only recently, quite distinct economic areas (Table 2). Major EU15 cities had a far higher MSW generation rate than the CEE cities in 2001, though the average annual growth in CEE cities was more than double that of cities in EU15 countries between 1995 and 2001. This trend suggests that waste amounts in the two areas will become similar in the future.

Table 2. MSW generation in EU cities

	MSW generated (kg/person/year)		Average annual change
	1995	2001	
Old member states (EU15)	466	519	+1.8%
New member states and applicant countries[a]	287	369	+4.3%

[a] Bulgaria, Romania.

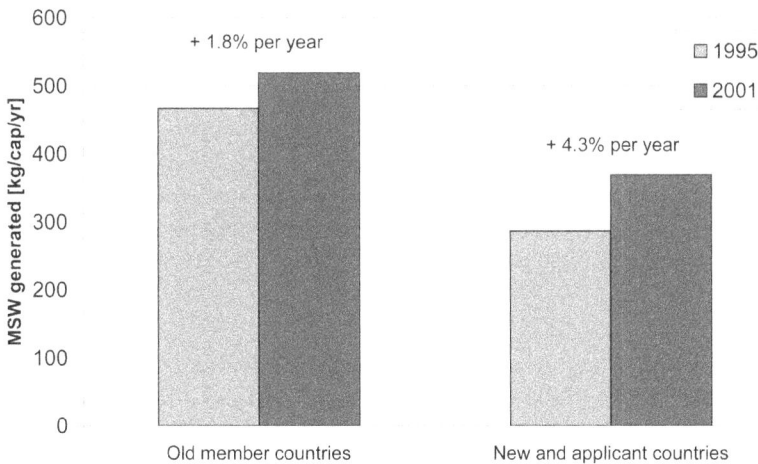

Figure 2. MSW growth rates in selected European countries

All European countries faced increasing quantities of MSW over this period. At the bottom of the scale, we found a growth of 1–8 per cent over the last 6 years. However, the highest increases, ranging from 20 per cent to more than 50 per cent, took place in Slovakia, Poland and Spain – all countries with rapidly growing economies (Figure 2).

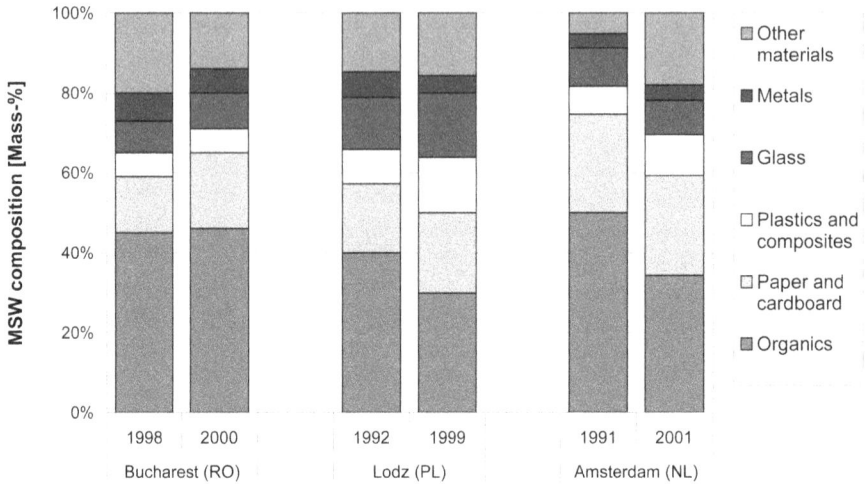

Figure 3. MSW composition trends in selected European cities

Composition trends

Poor availability of reliable sorting analyses mean that it was possible to document changes in MSW composition over a longer period in only a few cases; data from three cities are reported in Figure 3. The total percentage of packaging-related materials (paper/cardboard, plastics and composites, glass and metals) appears to be increasing, while the mass percentage (but not inevitably also the amount generated per person) of the organic fraction is tending to decrease.

Paper and cardboard

A twofold increase in the amount of paper and cardboard generated per person was observed between CEE and EU15 cities. The rate in CEE cities was 40–80 kg/person per year (mean: 56 kg/person per year), but, in most of the EU15 cities, it was 90–140 kg/person per year (mean: 113 kg/person per year).

Organic waste

No such significant differences were found with regard to organic waste. Despite considerable differences in the amount of MSW generated per person, 120–180 kg/person per year of organic waste was generated in nearly three-quarters of the data sets. One discrepancy was observed – considerably higher generation rates were reported for Spanish and Greek cities.

Plastics and composites

In terms of mass percentage, similar amounts of plastics and composites were documented; 10–15 per cent of total MSW generation for around two-thirds of the cities. Much lower values were obtained for cities with a low income (e.g., Bucharest and Baltic cities) as well as from results of sorting analyses in Polish cities from the early 1990s. These findings are linked to the later introduction of plastics as the main packaging material in these cities.

Other fractions

No significant prosperity-related variations were observed with regard to the generation of glass and metal wastes.

Influencing factors

The trends for total MSW generation and paper/cardboard appear to confirm the assumed relationship to the general level of prosperity. To quantify this impact more accurately, the investigated indicators were subjected to statistical analyses such as correlation and regression analyses or cluster analyses. Over 500 data sets, each referring to a city in a certain year between 1980 and 2001, were used to evaluate the impacts of the potential indicators at both city and national levels. This made it possible to consider regional peculiarities as well as changes in time.

To eliminate highly different socio-economic conditions between European cities, each city in a determined year was assigned to one of three groups depending on its 'prosperity level'. This was defined according to GDP, employment sector and infant mortality [3]. This allowed the development of a set of regression equations forming an estimation model for MSW generation. The significant indicators identified are shown in Table 3 and discussed below.

Table 3. Factors influencing MSW generation

Model parameter	Level of prosperity	
	Low[a]	High[a]
GDP[a]		+[b]
Infant mortality	−[b]	−
Population aged 15–59 years	+	
Household size	−	
Life expectancy	+	

[a] GDP below/above approximately US$13,800 purchasing power parity (1995).

[b] National indicator.

GDP

This commonly used indicator proved to be a significant factor in cities with high prosperity, but not in cities with a lower economic output. This is because the high regional income inequality in CEE countries causes a big gap between mean values, which are usually available, compared with the clearly lower median values – a more meaningful, but rarely observed indicator for social well-being and living standards.

Health indicators

Although only one previous study [4] has used infant mortality and life expectancy as parameters to indicate MSW generation, they showed a remarkable ability to serve as an additional or alternative variable for GDP. Their advantages include good availability and high quality of data.

Age structure

Age structure has also been undervalued as a possible predictor for MSW generation. The greater the percentage of the population aged 15–59 years, the more employed persons there are; this implies a subsequent rise in overall consumption and waste generation. The positive relationship between the percentage of this middle-age group and MSW generation was recently confirmed for German cities [5]. The parameter has also been shown to affect economic output and consumption [6].

It seems more than a coincidence that the three countries with the highest waste increase between 1995 and 2001 (Figure 2) also had the highest percentage of people in this age group (Spain and Poland: 64.2 per cent; Slovakia: 65.1 per cent (2000)) as well as the highest increase in this period (Poland and Slovakia: +2.8 per cent; Spain: +1.2 per cent) of all evaluated countries. Comparatively, Germany faced the biggest decrease (−1.7 per cent).

Household size

Household size is a standard indicator for the level of poverty. Our study confirmed a previous finding [7] that there is a significantly negative relationship between average household size and MSW generation.

How to predict future waste quantities

It is always difficult to decide the best way to deal with uncertainties. In planning practice, underlying assumptions or expectations concerning forecasts are, of necessity, not well defined; therefore, they are not transparent and are frequently overlooked. Although exact waste forecasts cannot be achieved, improvements in forecast accuracy are possible. However, they are often limited by:

- failure to consider social and economic trends related to the region;
- lack of reliable forecasts with regard to these trends.

The approach applied in this study attempts to mitigate these problems. First, the relationship between the set of indicators is well known, and it is assessed statistically on the basis of an investigated period of 22 years. Secondly, the model is based on prioritised parameters represented by factors with a high inertia such as health or demographic indicators. This means that they are easier to predict than highly volatile indicators such as GDP. Forecasts of this economic measure have an early expiry date, whereas forecasts for health or demographic indicators (e.g., those updated regularly by the United Nations) are available at a national level up to the year 2050 [8].

Our MSW generation model was used to develop the final forecasting model. An interpolation method was used to consider long-term changes between prosperity levels. Testing the accuracy of forecasts is crucial. Because the real error of a forecast over 10 years cannot be tested for practical reasons, the forecast accuracy was tested by ex-post forecasting. Accordingly, fictitious forecasts were made on the basis of the underlying independent variables for a historic year. The estimated values were then compared with the real value in this ex-post predicted year. The developed model leads to a median relative error of 5.3 per cent for all time series of 5–10 years and 7.7 per cent for time series of 11–22 years; these errors are related to the error at the end of the time series. The more relevant estimation error of the annual growth rate (derived thereof) amounts to approximately 0.6 per cent per year.

Underlying trends

All relevant indicators but one foster the assumption of a further growth in MSW generation per person in European cities.

- Economic growth – long-term estimates from the World Bank [9] suggest that moderate GDP growth rates for EU15 (+2.3 per cent) and higher ones for CEE countries (+3.3 per cent) can be expected.

- Health indicators and household size will follow the previously observed trend of increasing welfare standards and shrinking household sizes.

- Rising urbanisation – an increasing share of the population will live in cities, though the population in existing major cities will remain constant.

A contrary, retarding effect on further increases in waste generation could arise from the decreasing share of the population in the mean age group. Following a peak in most European countries between 2000 and 2005, demographic forecasts suggest a steady decline back to the level of the 1980s or even the 1970s.

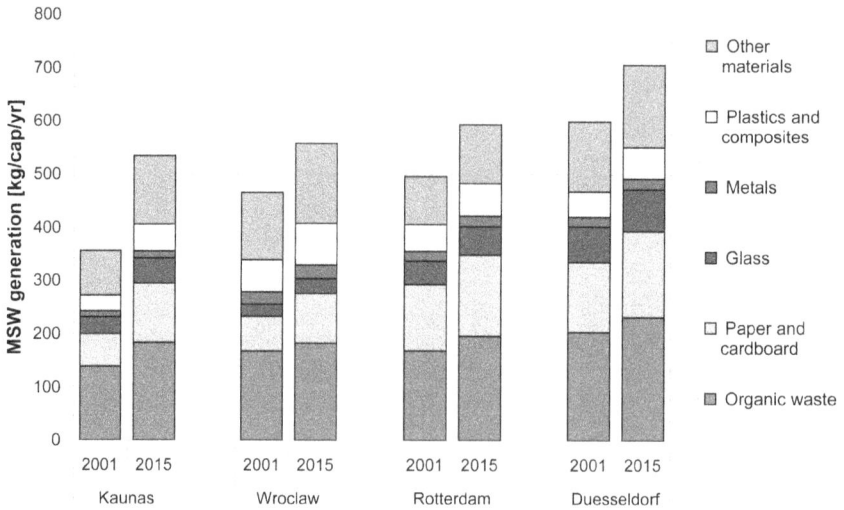

Figure 4. Forecast results for selected European cities

Forecasts

Due to regional trends, global forecasts for cities in Europe are neither particularly useful nor helpful. Figure 4 shows forecasts for four European cities – two from the EU15 and two from the accession countries.

Although not guaranteed valid for all cities, the following trends can be deduced:

- further rise in MSW generation;

- equalisation of MSW generation rates due to stronger growth in CEE countries;

- disproportionate increase in the generation of waste paper and cardboard;

- rate of increase of organic waste generation less than that for total MSW.

Application

The use of improved waste-generation forecasts enables more adequate dimensioning of the waste collection infrastructure and treatment facilities, thus helping to avoid expensive over- or undercapacity. Compliance with the recycling and recovery targets of the EU packaging directive serves as an example. These targets are defined as a percentage of the total future generation by material. There is thus no direct indication of the total volume that will need to be processed. Use of the model to predict the volumes of different materials that could be generated would help to reduce planning uncertainties when designing separate collection systems and recycling facilities to fulfil the prescribed values.

Application of the presented model allows the compilation of regionally adapted forecasts based on the user input of a small set of city-related indicators. Missing data problems are avoided by proposing default values from an extensive background database. To verify its practicability, the software tool has been tested in five European cities from regions with rapidly growing economies.

Conclusion

Bringing the zero waste philosophy down to earth requires a framework about quantity and quality of future wastes to be reduced. The model presented here is an attempt to improve existing planning practices without broader considerations. Dynamics of waste generation and composition of MSW are estimated. Combined with recycling scenarios, the needed capacities of facilities can be assessed more accurately. But, as mentioned before, this is only an attempt at improvements. Much has to be done to enhance knowledge about the characteristics of waste fractions and – of course – waste producers. Only recycling, an element of zero waste, is addressed in this chapter. To go beyond towards eco-design and waste prevention, more comprehensive data and refined methods are needed as framework for an all-embracing zero waste.

Notes

More about LCA-IWM
The model outlined in this chapter was implemented in a platform-independent, Java-based decision-support tool. The software, related handbook and project report may be downloaded at www.lca-iwm.net.

Acknowledgements

Data collection and inspection were carried out in cooperation with the following LCA–IWM project partners: Darmstadt University of Technology, Germany; Democritus University of Thrace, Greece; Syncera De Straat, The Netherlands; novaTec, Luxembourg; Universitat Rovira i Virgili, Spain; and Wrocław University of Technology, Poland.

References

[1]   Beigl, P., Lebersorger, S. and Salhofer, S. 'Modelling Municipal Solid Waste Generation: A Review', *Waste Management*, 2008, 28(1), 200–14.

[2]   Fischer, C. and Crowe, M. *Household and Municipal Waste: Comparability of Data in EEA Member Countries*. Topic Report No. 3/2000 (Copenhagen: European Environment Agency, 2000).

[3]   Beigl, P., Wassermann, G., Schneider, F. and Salhofer, S. 'Forecasting Municipal Solid Waste Generation in Major European Cities' in Pahl-Wostl, C., Schmidt, S., Jakeman, T., eds. *iEMSs 2004 International Congress: Complexity and Integrated Resources Management* (Osnabrück: 2004). www.iemss.org/iemss2004/pdf/regional/beigfore.pdf.

[4]   Bogner, J., Rathje, W., Tani, M. and Minko, O. 'Discards as Measures of Urban Metabolism: The Value of Rubbish', Paper presented at an international symposium on urban metabolism, University of Michigan Environment Dynamics Project, Kobe, Japan, 1993.

[5]   Sircar, R., Ewert, F. and Bohn, U. 'Ganzheitliche Prognose von Siedlungsabfällen', *Müll und Abfall*, 2003, 1, 7–11.

[6]   Lindh, T. 'Demography as a Forecasting Tool', *Futures*, 2003, 35, 37–48.

[7]   Dennison, G.J., Dodd, V.A. and Whelan, B. 'A Socio-Economic Based Survey of Household Waste Characteristics in the City of Dublin, Ireland. II. Waste Quantities', *Resources, Conservation and Recycling*, 1996, 17, 245–57.

[8]   Beigl, P., Gamarra, P. and Linzner, R. 'Waste Forecasts Without "Rule of Thumb": Improving Decision Support for Waste Generation Estimations', SARDINIA 2005 – Tenth International Waste Management and Landfill Symposium, Paper no. 251 (in press), CISA Environmental Sanitary Engineering Centre, Cagliari, 2005.

[9]   International Bank for Reconstruction and Development/World Bank (IBRD). *Global Economic Prospects and the Developing Countries* (Washington, DC: World Bank, 2003).

15

Paper recycling, recovery and upcycling: a cornerstone for sustainable use of renewable resources

Ranjit Singh Baxi

J & H Sales International Ltd, UK

This chapter discusses a key dimension of 'WASTEnomics' – that of the recycling, recovery and upcycling of recovered fibre (paper waste) – which ultimately helps to minimise waste and turn the liabilities of waste into assets.

Introduction

The first piece of paper as we know it was produced from rags in AD 105 by Ts'ai Luin, who was part of the Eastern Han Court of the Chinese Emperor Ho Ti. Today the production of paper is a global industry and accounts for 40 per cent of all wood-based products. This equates to millions of trees that are cut down every year, and several hundred million tonnes of greenhouse emissions into the atmosphere as a result. Additionally, 98 tonnes of other resources are required to produce 1 tonne of paper.

It is clear that the effect of the paper industry on forests is staggering. For example, Canada alone cuts down 5 acres (2.02 hectares) of forest every minute, much of that going directly and indirectly to feed pulp and paper mills.

Furthermore, the world faces the challenge of managing the millions of tonnes of 'post-consumer paper' generated annually. In recent years, up to 30 per cent of such paper was recovered for recycling, the remainder being disposed of – primarily in landfills.

The annual worldwide generation and consumption of paper continue to increase, and recycling alone will not necessarily reduce the amount of paper sent to landfills. All major sectors of society – whether manufacturers, businesses, government organisations, or private residents – must also reduce the amounts of paper used. Reducing our consumption of paper not only benefits the environment, but also often improves the bottom line and operating efficiency of a business.

Paper and fibre recovery is the practice of collecting and reconstituting paper to produce recycled paper or some other product, such as cellulose insulation. Paper and fibre recovery is accomplished in many ways. The most visible method is residential recycling programmes, which collect a variety of paper types generated by households. However, the majority of recovered fibre and paper originates from the business and industrial sectors.

About 20 per cent of all paper recovered is 'pre-consumer' paper (that which never reaches consumers). Pre-consumer paper is generated by sources such as converters (businesses that take rolls of paper and convert them into finished products) and printers (e.g. misprints and overruns).

Given the global annual consumption of about 340 million tonnes of paper and assuming that we need about 13 trees to produce the pulp required to make 1 tonne of paper, a colossal number of trees must be cut down. However, the contribution of the recovered fibre industry in producing nearly 140 million tonnes of paper needs to be taken into account – over a billion trees are saved as a result of the recovered fibre industry and our use and consumption of recovered fibre! This is indeed sustainable use of our renewable resources (Figure 1).

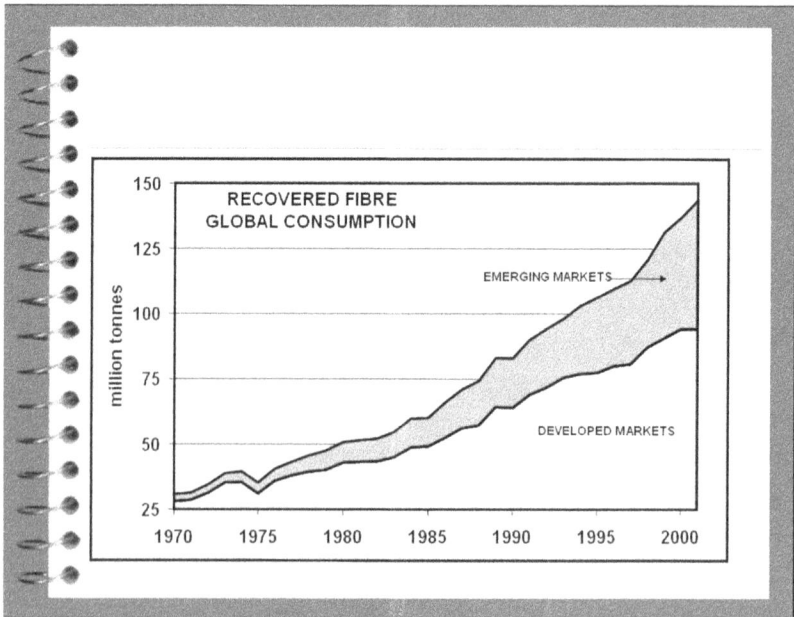

Figure 1. Global consumption of recovered fibre

Box 1. Feed stock for making recycled paper

There are three categories of paper that can be used as feed stock for making recycled paper: *mill broke*, *pre-consumer waste* and *post-consumer waste*.

Mill broke is waste material recovered from inside a paper mill during paper-making process. This includes paper trimmings. Paper makers have always reused their broke. Many people do not consider paper made from mill broke to be genuine recycled paper.

Pre-consumer waste refers to waste paper that has been converted and perhaps printed but which has been discarded prior to reaching the consumer. This includes printer off-cuts, envelope trimmings and rejected stocks.

Post-consumer waste is paper that is recovered after it has been used as a consumer item. It includes waste paper from offices and homes, old newspapers, old telephone directories and packaging.

Source: 'Debunking the Myths of Recycled Paper', *Directions*, February/March 1992.

Paper recycling: a cornerstone for sustainable use of renewable resources

In our modern society, we use millions of tonnes of paper. The collection and recycling of used paper makes sound economic and environmental sense. Most paper products, once used, and provided they are separately collected, can begin a new life as a secondary raw material – several times over. As such, the recycling and recovery of paper are a perfect example of the paper industry's sustainable use of renewable resources. Paper recycling and recovery make a vital contribution to our planet's sustainable development, economically as well as environmentally, creating jobs and saving energy and natural resources. Given that recovered paper's share of the raw materials used for paper making is only 42 per cent (wood pulp accounts for 43 per cent with the rest being non-fibrous materials and other pulps), it is clear that the current level of paper production could not be maintained without recycling.

Wood is the most important renewable resource in Europe, and is used both as a raw material and as fuel. Based on wood, recovered paper forms a source of renewable raw material, and could, in principle, be considered a renewable energy source too. Reasons for this are well justified, such as increasing the energy self-sufficiency of the European Union (EU) and combating climate change. The starting point in analysing the carbon footprint of the paper industry is the capacity of forests – the origin of paper and board raw materials – to bind CO_2 and contribute to the mitigation of climate change. If forests are managed sustainably, trees are renewable and recycle carbon from the atmosphere; therefore, they have a neutral effect on the amount of atmospheric CO_2.

Box 2. The 10 toes of paper's carbon footprint

1. Carbon sequestration in forests – sustainable forest management (SFM) allows the stocks of carbon in forests to stay neutral or even improve in time.

2. Carbon in forest products – a product contains biomass carbon and as long as it is in use, it will keep this biomass carbon from the atmosphere.

3. Greenhouse gas emissions from forest-product-manufacturing facilities – these emissions come from fossil fuel combustion at manufacturing facilities that produce forest products, including primary manufacturers and final manufacturing facilities.

4. Greenhouse gas emissions associated with producing fibre – for virgin fibre, this includes forest management and harvesting; for recovered fibre, it includes collection, sorting and processing of recovered paper before it enters the recycling process.

5. Greenhouse gas emissions associated with producing other raw materials and fuels – these emissions are generated during the manufacture of fuels and non-wood-based raw materials (e.g. chemicals and additives) used in manufacturing forest products, and they are also direct emissions and emissions associated with purchased electricity to manufacture these raw materials.

6. Greenhouse gas emissions associated with purchased electricity, steam, heat and hot and cold water – these emissions are associated with purchased electricity, steam and heat used at facilities that manufacture forest products, including chip mills, pulp mills, paper and paperboard mills, and final manufacturing facilities (e.g. box plants). This includes electricity for pollution control equipment used to treat manufacturing-derived wastes and emissions.

7. Transport-related greenhouse gas emissions – these emissions are associated with transporting raw materials and products along the value chain. It includes emissions from transporting wood, recovered fibre, other raw materials, intermediate products, final products and used products.

8. Emissions associated with product use – these emissions occur when a product is used. These are very unusual for forest products, and this is a key advantage of forest products over, for example, electronic media.

9. Emissions associated with product end-of-life – these emissions occur after a product is used. They consist primarily of methane (CH_4) resulting from the anaerobic decomposition of forest products in landfills.

10. Avoided emissions and offsets – these emissions do not occur (i.e. are avoided) because of an attribute of the product or an activity of the company making the product.

Source: Confederation of European Paper Industries, *Waste Management World*, 13 November 2007.

Paper and fibre recovery

Approximately half of all recovered paper and fibre is sourced from industry and businesses in the form of converting losses – such as cuttings and shavings – packaging, and unsold newspapers and magazines. Well over a third comes from householders, paper collection and recovery representing one of the most direct ways in which members of the public can participate in the environmental and social goals of recycling. Almost any household paper can be recycled, including used newspapers, cardboard, packaging, stationery, direct mail, magazines, catalogues, greeting cards, old telephone directories and wrapping paper.

In many instances, recovered paper and board account for more than half the volume of raw material used in the production of new products. High recycling rates are achieved in many countries. In Europe, for example, the recovered paper industry met the voluntary target of a 56 per cent recycling rate by 2005 (with a new target of 66 per cent by 2010). Meanwhile, developing countries are looking to develop their own collection infrastructures in order to reduce their dependence on imports for feeding domestic paper and board production plants.

Box 3. Main sources and applications of recovered paper

Recovered paper originates from households (40 per cent), commercial and industrial sources (50 per cent) and offices (10 per cent). In 2005, the collection from European households and offices grew at the rate of 5 per cent, and these sources, not fully tapped yet, offer the greatest scope for increasing paper recycling. Applications for recovered paper are approximately two-thirds of the material currently being used to produce corrugated board and newsprint.

Source: Confederation of European Paper Industries.

Already more than six centuries old, paper recycling has grown substantially during the last few years and continues to expand. Many packaging materials, newsprint and tissues are made wholly or in part from recycled fibre. The recovered paper industry collects material, sorts and segregates it into various types, and processes it for ease of handling, transport and subsequent repulping. It uses many different kinds of machinery to fulfil these functions, including massive baling machines that apply huge forces to convert recovered paper and board into dense 'cubes', which are easier and more cost-effective to transport to consumers, especially in containers for export.

If it were not for the efforts of the world's paper-recycling industry, a large proportion of unwanted paper – in the form of, for example, off-cuts from printing companies or newspapers from households – would end up in the waste stream and thus in landfills. The industrial generators of unwanted paper and board or the taxpayer would have to pay for its disposal.

All paper making is based on cellulose: recovered cellulose fibres are parted in the paper industry's pulping process, sophisticated equipment and various chemicals being used to clean and condition the pulp so that the end product conforms to the strictest standards of hygiene and cleanliness. Successive repulping tends to lower the quality of the fibre until, in theory, fibre collapse occurs; it has been estimated that paper can be recycled on average four to six times, depending to some extent on the paper grade in question. Therefore, virgin primary fibres are added to maintain strength and other qualities, and recycling processes can provide for damaged fibres to be removed.

Box 4. Economic and environmental savings from recycling and recovery

For every tonne of paper used for recycling, the savings are as follows:

- at least 30,000 litres of water
- 3,000–4,000 kW of electricity (enough for an average three-bedroom house for 1 year)
- 95 per cent of air pollution.

Source: WasteOnline Paper Recycling Sheet 2007.

In terms of environmental pollution and energy consumption, recovered paper compares favourably with the production of wood-based pulp made by chemical or mechanical means. As fresh wood fibres are needed to guarantee paper recycling, recovered paper and forest products complement each other both ecologically and economically.

For paper recycling and recovery to realise greater benefits, the generators of paper have to see the advantages of making their material available for collection instead of allowing it to be dumped or destroyed. At the same time, the paper-recycling industry has to cover its collection and processing costs, while industrial users of secondary material have to be able to manufacture marketable and competitive products. One of the key issues for the recycling industry over recent years has been to create additional demand for products containing high levels of recycled fibre.

Box 5. Analysis of demand from major global markets

Increasing globalisation and the rapid growth of the paper and board production base in key developing countries have had a huge impact on movements of recovered paper and board. Worldwide trade currently amounts to more than 175 million tonnes per annum, offering one of the most glaring proofs that the recycling industry is engaged in the production of a product rather than a waste (see Figure 2 below). It is worth noting, for example, that the United Nations Basel Convention prohibits or severely restricts the transport of waste between countries, while the People's Republic of China bans waste imports. However, China welcomes the import of millions of tonnes of recovered paper and board every year – from Europe, North America and Japan among others – to feed its growing domestic production base. Indeed, it imported more than 25 million tonnes of recovered paper in 2007.

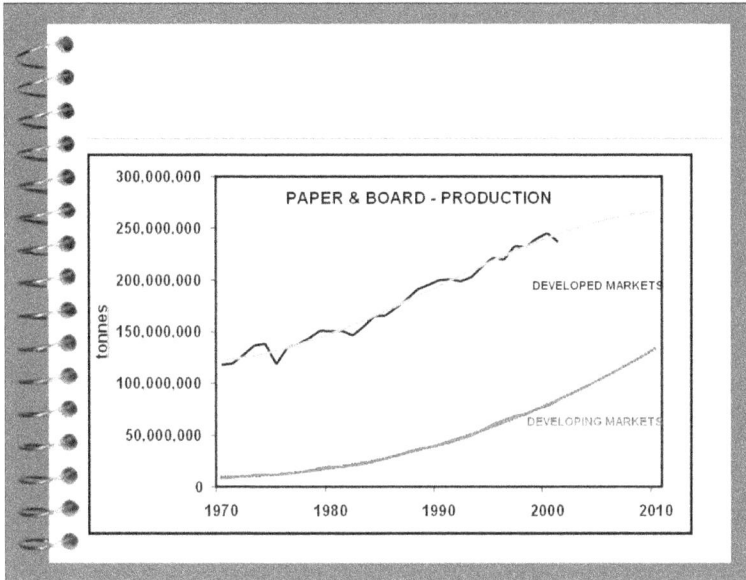

Figure 2. Paper and board production

Europe

In 2007, 13 different sectors in the paper value chain pledged their support for the new European Declaration on Paper Recycling covering all paper and board products, and aiming to push the European paper recycling rate to 66 per cent by 2010.[1]

A total of 58.2 million tonnes of paper and board was recycled in Europe in 2006; an increase of 5.7 million tonnes (or a 9.8 per cent increase) since 2004, the base year for the target. Recycling of paper is a significant part of the paper-manufacturing process in Europe, but it is also a large industry in its own right, with links to a number of sectors in the global economy.

The European Declaration focuses on complementary actions by all of the sectors involved, and gives priority to the prevention of waste, improving the recyclability of paper and board products by eco-design, and further improving the quality of recovered paper available for recycling.

Reaching the target of 66 per cent would mean that some 2 tonnes of paper are recycled in Europe every second. Europe is already the global leader in paper recycling with a regional rate of 63.4 per cent, higher than either Asia or the North America. The industries along the paper value chain have nevertheless raised the bar further, aiming to reach the 66 per cent target by 2010.

---

1   The Monitoring Report 2006 gives an overview of the sectors' achievements in 2006, including the latest figures of the European paper-recycling rate.

India

Current demand for paper in India is 7.2 million tonnes, while the production level is 6.6 million tonnes. The deficit of 0.6 million tonnes mainly comprises newsprint imports. With increased consumption and rapidly growing readership, new capacities will come on stream that will further increase the hunger for fibre in India. Moreover, demand for finished paper will outgrow supply, forcing increased importation of finished paper alongside increasing fibre imports.

Newsprint demand in India is projected to grow at 5.6 per cent annually until 2020, compared to northern Europe, where it is growing at a modest 1.2 per cent, while Japan and North America will see an actual drop of 0.5 per cent. With a projected readership of 750 million by 2010, the demand in India is projected to exceed the supply by 1.0–1.3 million tonnes, resulting in even greater import of newsprint.

China

China is deficient in fibre supply and will remain highly dependent on imports of recovered fibre. The next 3 years will see a significant increase in fibre imports by both China and India as well as other Asian countries.

China's paper industry is expanding at an amazing rate, leading to over-capacity and resulting in increased paper exports from China. The expansion of China's paper industry by modern, large-scale, high-speed, and competitive production methods will make its exports globally competitive.

China is the second largest producer of paper and board (behind the USA), having experienced growth of nearly 12 per cent per annum over the past 5 years (compared to just 5 per cent per annum for the 5 years before that) (Figure 3).

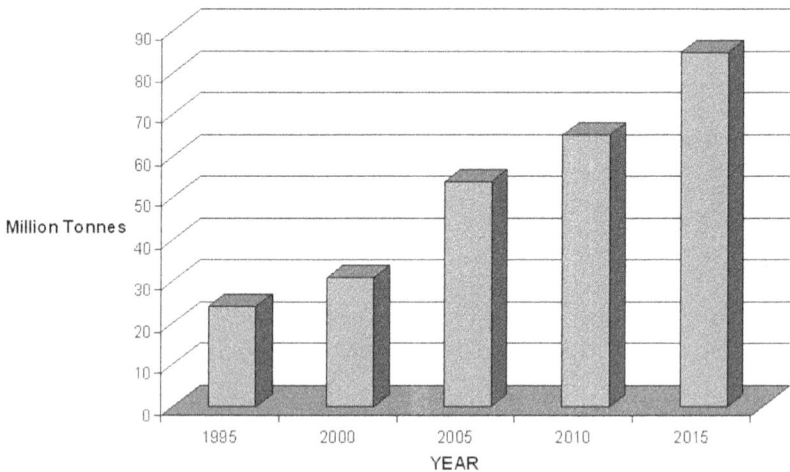

Figure 3. China: production of paper and board, 1995–2015

As a result, mills and machines are shutting down in both Europe and North America, making fibre export a necessity from there. At the same time, we will see more machines in the future being repositioned mainly in India from Europe and North America to satisfy increasing and growing local demand.

Given such scenarios, there is a clear need for India and China to increase domestic recovery levels, as imports alone cannot satisfy total fibre demand. China's recovery level needs to go beyond 50 per cent.

Paper upcycling?

The production of paper requires bleaching. We can recycle the paper, which becomes recycled paper. However, when this paper can no longer be recycled, it will end up in a landfill, where the chlorine that was used to bleach the original paper will be leached out.

This form of recycling – generally known as 'down-cycling' because of successive downgrading – merely postpones the consequences of the problem. At some stage, the paper can no longer be recycled due to the continual degrading. Given the process of resource depletion and continual degrading, it is clear that we need to address the process of down-cycling. One approach is the process of 'upcycling'[2], which completely reuses materials in ways that do not degrade their quality.

The upcycling approach is considered 'eco-effective', given that it does not degrade the quality and its subsequent disposal. On the other hand, conventional recycling is considered as 'eco-efficient' given the continual degrading until it reaches a situation where it cannot be recycled. Eco-effective companies using such an upcycling approach can potentially lower their raw material costs and generate more sustainable returns.

The problem we should solve is how to design products that have only biological constituents so that they can be recycled continually. For example, books are products that combine biological constituents (paper) and artificial constituents (adhesives) that prevent them from being fully recycled.

---

2   Upcycling is the practice of taking something that is disposable and transforming it into something of greater use and value. The term was coined by William McDonough and Michael Braungart, authors of *Cradle to Cradle: Remaking the Way We Make Things* (New York: North Point Press, 2002).

Box 6. Example of an upcycle system

An exemplary upcycle system involves the creation of a closed-loop publishing cycle, in which the publisher replaces traditional paper with synthetic paper. The durable nature of synthetic paper enables the publisher to recirculate, de-ink, or upcycle the paper, depending on the nature of the designed publishing cycle. The upcycle system involves four primary partners: publisher, consumer, carrier, and upcycler.

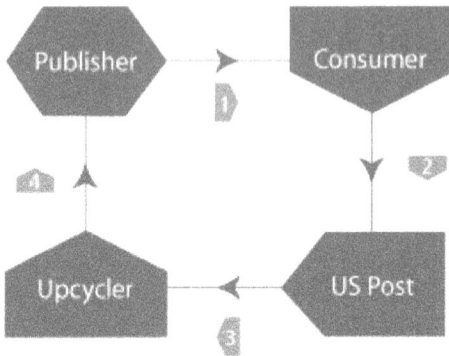

Source: The Upcycle System.

Conclusion

The world faces the challenge of managing millions of tonnes of pre-consumer and post-consumer paper generated annually. With the growing annual global generation and consumption of paper, recycling alone will not necessarily reduce the amount of paper sent to landfills. When this problem is coupled with the expected rise in the world population from 6 to 9 billion by 2050 and the annual migration of 33 million people from the rural to the urban environment (and its associated modern offices and paper usage), it is clear that the problem of collecting and recycling paper waste will become significantly more important and require serious consideration.

We must also take into account the contribution of the recovered fibre industry in collecting over 175 million tonnes of fibre that is then used by the paper industry to produce nearly 140 million tonnes of paper. This translates into saving the equivalent of over a billion trees (and the resulting avoidance of greenhouse gas emissions) as a result of the recovered fibre industry and our use and consumption of recovered fibre! This is indeed sustainable use of our renewable resources. More collection, recovery and high-quality recycling are the only way forward if we are to contribute to safeguarding not only our environment but also that of our children in the future.

16
Zero waste – the achievable dream

Stewart Anthony
Middlesex University, London, UK

Jane Fiona Cumming
Article 13, London, UK

This chapter introduces a key dimension of 'WASTEnomics' – that of zero waste, and addressing each key strategy of the waste hierarchy and ensuring effective implementation of these strategies throughout the organisation.

Introduction

In the general media clamour about global warming, political sound bites and scientific concern, the associated environmental issues which are imposing themselves on this planet are in danger of being ignored or at best seen as being of less importance. The reality, of course, is that not only are some of the environmental pressures on the planet just as important as global climate change, but they are also inextricably linked to it.

Eliminating waste from all the production and service processes in an organisation and then extending that to the elimination of waste from household activities is not only an objective that is possibly more achievable than reduction of energy consumption by 60 per cent by 2050 (UK Climate Change Bill 2007), but would also significantly contribute to reduction of global warming, reduction of air pollutants, protection of forests and other ecological areas, and improvement of many other environmental aspects where over-consumption of resources and dumping of wastes results in gross destruction of land, a non-renewable resource.

The concept of zero waste has been developed in recent years as a means of establishing a target that would have major environmental benefits and also stimulate creative thinking. It was hoped this would be achieved not only through technical changes and improvements, such as new designs, technologies and materials, but also through addressing business responsibilities and then communicating with and training staff to operate and think in a way to lead to the complete elimination of waste in all its forms.

This chapter will critically examine the issues which face the move towards a zero-waste economy and way of life, and will relate these to both corporate governance issues and the problem of communicating with and training staff to change their attitudes and behaviours in order to achieve zero waste.

The pressure to eliminate – and the pressure to go on producing – waste

Within the UK, the traditional method of disposing of waste has been to dump it in large holes in the ground. With some wastes, notably sewage and particularly hazardous materials, the sea has been the dustbin to which we have consigned these materials until pressure to cease became paramount. But the UK is running out of holes in the ground. Political and environmental pressures have reduced marine dumping, and for businesses and the general public the cost of dumping waste is beginning to hurt. Some parts of the UK are estimated to have only about five years of landfill space available, and the imposition in 1996 of the landfill tax, which has been escalating ever since, is starting to make businesses realise that throwing waste away is affecting their profits. Media reports, resulting from political 'kite-flying', that householders may have to 'pay as you throw' are causing widespread anger, frustration, or concern among householders, especially as they think the waste problem will end in costing them more!

In other European countries, landfill pressures are not so considerable, but there is still need to reduce waste. Pressure to reduce waste also comes from the increasing pressure on resources. Oil prices, which have been rising steadily, show no sign of falling, and since oil is the major raw material of plastics, there is a consequent pressure there. Metal prices are also increasing due to resource demand, especially from infrastructure projects in China and India.

The impact on global climate change of waste, as it gives off methane, carbon dioxide and other global warming gases, is an additional pressure to reduce waste.

However, there is clearly a tension between waste elimination and the desire to continue consumption of the goods, materials and services which produce waste. Since the early 1990s, very few countries have not adopted or participated in the market-driven economy that now drives economic development worldwide. The problem is that waste is not part of that market economy. Waste is a by-product of the market economy that hardly anyone wants, and products made from waste materials are (actually or perceived to be) of inferior quality and performance, too expensive, or just not fashionable. In a system in which dumping of goods is far less costly than recycling or remanufacturing, there will always be a struggle to eliminate waste totally from that system.

It is also in this context that the message of 'reduction' has simply failed to get through either to the businesses and organisations that produce the goods and materials, or to the public who are the ultimate consumers (or even the designers of products or the manufacturing process for these products!). The term which has been communicated most effectively is 'recycling'. In fact, recycling is the fourth or fifth step in the waste hierarchy, and in terms of material and energy consumption, it is a much less realistic choice than redesign, reduction or reuse. One reason why recycling appears to have been communicated as 'the way to reduce waste' is that it does allow the market economy to function and flourish. We, as consumers, can keep on purchasing goods, materials and services just so long as when we have finished with them, we prefer to recycle them rather than throw them into the dustbin. Interestingly, recent experience with many small business enterprises in north London resulted in encounters with businesses who freely use the term 'recycling' in place of 'waste management' (Box 1).

Box 1. Case study – cardboard boxes

In north London, a wholesaler of ladies' fashions to several major high-street chains demonstrated how to build up their cardboard box waste quantity. They would take in a range of clothes from different manufacturers all arriving in different-sized cardboard boxes. The clothes would be unpacked and hung prior to batching according to the retailers' requirements, whereupon they would be packed into large, new cardboard boxes for transport. Analysis of the case found that the transport company charged the wholesaler by the number of boxes, not by their size or weight; therefore, the wholesaler sent out goods only in large, new cardboard boxes.

The consequence of this was that the wholesaler was dumping cardboard boxes they could have reused and buying large cardboard boxes which were not always needed, the retailer was having to dump larger cardboard boxes than necessary, and the transport company was carrying quite a lot of air. When the Centre for Environment and Safety Management for Business (CESMB) worked with the company, they estimated that this was costing the company over £40,000 per year.

The wholesaler responded that 'our clients demand new cardboard boxes', though they admitted that they had never specifically discussed this with clients, and that it was more efficient to always use the same-size boxes, though they could not justify this.

Zero waste – the likely strategies

The achievement of zero waste will have a number of interrelated strategies, but underpinning them all will be the waste hierarchy (Figure 1).

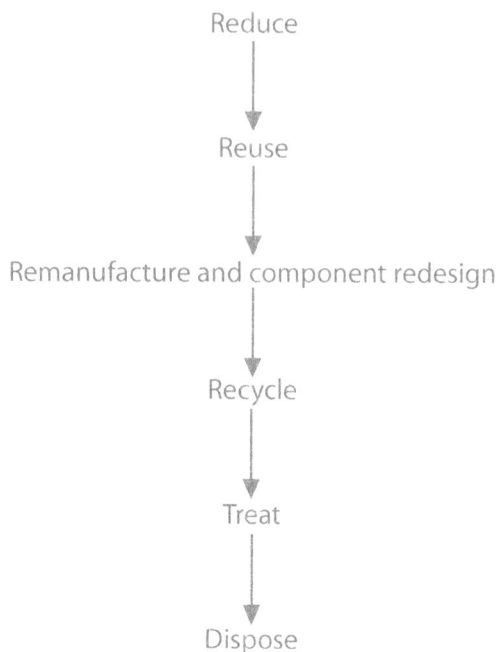

Reduce

↓

Reuse

↓

Remanufacture and component redesign

↓

Recycle

↓

Treat

↓

Dispose

Figure 1. The waste hierarchy

In this hierarchy, the most effective waste-elimination strategy is to reduce resource consumption.

In its simplest approach, this would clearly have a serious impact on the global economy and is generally an approach which all commercial organisations will resist. However, this simple approach and this simple resistance are not the way forward. Reduction of consumption should mean a move towards responsible consumption of products, materials, and services made from or using renewable and sustainably managed resources; that is, goods and materials designed to be reused again and again, rather than used once and dumped.

This approach means consumption of goods and materials designed to be able to be disassembled and partially or completely refurbished (cf. Interface) – so-called remanufacturing. Responsible consumption implies that any materials that cannot go through this 'naturalistic cycle' should be treated to release their energy to power electricity-generating systems or composted for use in horticulture or agriculture, replacing 'virgin' resources from elsewhere.

Box 2. Case study – reducing waste

A company in north London that sell wholesale electricity cable to utility companies and DIY chains identified the very high cost of their purchase, use and disposal of the wooden reels on which the cable was rolled. Each large reel cost the cable company £9, and each smaller reel cost £3. Depending on the amount of cable required by the customer they sent it on the appropriate sized reel. After use, the clients dumped the empty reels. The cable company worked with their clients and the transport company to institute a 'return reel scheme; for each large reel returned, the client was paid £3 and for each smaller reel, £1. In addition the transport company received £1 and £0.25, respectively, for each of the reels they brought back.

The winners in this scheme were clearly the cable reel company, the transport company and the clients, who all saved money, and the clients dramatically reduced their waste. The losers were the manufacturer of cable drums and presumably the timber company.

A critical component of any zero-waste strategy is collaboration throughout the supply chain (Box 2). In this responsible consumption strategy, there will clearly be winners – they are essentially the organisations who purchase less resources and pay less to dump less waste. However, there will also be losers – in the case study of Box 2, the loser is the manufacturer of the cable reels and possibly the timber company. Therefore, in a zero-waste strategy, communication is necessary not just to explain the benefits to the main players, but, much more difficult, to persuade the losers that they need different product or service strategies.

In the case of zero waste within the domestic waste management area, the identification of the losers is much more complex, but the advantages are very significant. A small project in Monmouthshire, UK, exemplifies the attempts being made to get to a zero-waste target and the importance of different communication strategies within that project (Box 3).

Box 3. Case study – the Zero Waste Village Project

The Zero Waste Village Project (ZWVP) was set up and operated by Monmouthshire Community Recycling. Some 84 per cent of the population used the original recycling scheme, but surveys suggested that there was potential to increase the range of recyclates collected, and the additional materials now collected include plastic bottles, rigid plastic food containers, and cardboard drinks cartons.

Residents of the ZWVP now have more kerbside collections than any other area of Wales. The final material (cardboard drinks cartons) was added on launch day of the project, along with St Arvans' own bring site and workplace recycling collections from local businesses.

Since the project was launched, the participation rate, recycling rate and overall view of waste management practices in St Arvans have improved. Ninety-five per cent of the houses in St Arvans participate in the kerbside recycling scheme, and 77 per cent of the waste produced in the ZWVP is diverted from landfill, exceeding some of the top European performers.

Operational expansion of the project is at its limits, and residents' participation is one of the highest in Wales, and probably the UK. Future attention to this project will centre on changing residents shopping habits to stop non recyclable products and packaging before they are even purchased.

A key aspect of the success of this project has been to engage with the communities through a wide range of communication strategies. This has included meetings in the villages with a range of resident groups, posters, meeting with individual residents and working with the schoolchildren to get them involved at an early stage.

Courtesy, Simon Anthony, Zero Waste Coordinator, Monmouthshire Community Recycling.

A zero-waste strategy which follows the waste hierarchy in Figure 1 will need to examine every step within that waste strategy from the top down. The reductionist approach and the reuse approach are most likely to face both commercial and personal hurdles: the desire of people to continue increasing their levels of consumption will be difficult to change and commercial organisations wedded to current unsustainable practices and processes will continue to encourage that consumption. It therefore seems reasonable that a more likely short-term change to zero waste would be achieved through examining the design and technological issues which can be altered to produce less waste.

The examination of current design issues should start from the premise that the product has to be reused, or has to be remanufactured, or, at worst, has to be recycled. Current design approaches to many everyday objects focus on the users' requirements, which include fashion; a zero-waste approach to design would need to focus on environmental requirements. In addition, the redesign should not just include the product but should also address the process design so that waste is designed out of the process from the word go.

Box 4. Recycling bins

Article 13 have recently been delivering a range of environmental training programmes to representatives of major UK companies. One of the interesting disparities between different companies was the degree of success in recycling office waste. The less successful organisations had simply installed bins for paper, glass, card, and plastic in the offices along with appropriate signage, and the result was very low rates of recycling. Other organisations had removed the general waste bins when the recycling bins were installed, and the result was complete uproar for about a week until everyone got used to the new 'way of doing things', but now recycling is the norm.

Beyond the design and technological issues, we must address behavioural patterns that will need to change in order to realise the full potential for waste elimination at each stage (Box 4). It is perhaps at the stage of trying to engender behavioural change that the greatest challenge will be faced, and where the responsibility of corporate governance, and the role of communication and training will be most required.

Clearly, communication about waste elimination is important, but it is only with the reality of having to put it into practice that people realise just how it will affect them, and that is where training, organisational culture, incentives and management commitment are all integral to the entire waste-elimination strategy.

Zero waste as a component of corporate social responsibility (CSR)

In businesses of all kinds, government bodies at all levels and academic institutions of every type, in NGOs and specialist consultancies, one can hear the terms 'corporate social responsibility' (CSR), 'sustainability' and 'sustainable development' being used with increasing frequency. The concept of CSR has been around for a number of years and has been defined in a number of different ways, but all the definitions have one thing in common – the idea that organisations have an obligation to consider the interests of everyone who is affected by their activities. In short, it is about businesses responding to the challenge of sustainable development (environmental, social, ethical and economic agendas).

But is CSR relevant to zero waste and is the concept of zero waste relevant to CSR strategies? In our experience with clients, the key to unlocking a CSR approach is to find practical programmes and concepts that can act as a starting point to get everyone engaged – from employees to the board. Focusing on waste brings increasingly personal values together with those of the organisation and embodies the concept of sustainable development in a very practical, visible and win–win situation. What board of any organisation would turn down the opportunity to make their business more resource efficient? The environment and climate change is the third issue in staff surveys for employees (after pay and benefits).

Bring the two together, without even understanding the concept of sustainable consumption, production or the science of the waste cycle, and the picture is complete for engagement with all stakeholders on a level playing field. It is about taking individual action for a shared vision. In one recent case, employees were asked about this as a concept, and they asked whether the savings made could be used in local community projects.

Recognising waste management as a business benefit issue

Just-in-time management, efficient consumer response, and lean manufacturing all have one thing in common – reducing inventory storage. So it is with zero waste. It is about using just what is needed, no more and no less, and rethinking processes and behaviours to eliminate the concept of waste. One of the major problems that businesses have with managing their current waste is that they tend to cost it as only the price they pay their waste contractor to remove it from their premises.

In reality, of course, the cost paid to the waste contractor is the tip of the waste-cost iceberg. The remainder of the iceberg includes buying the wasted resources in the first place, including staff time, and the costs of processing, storage, transport, etc.

Like energy, waste is a serious issue that can benefit businesses. No longer should business say, as they tended to say in the 1980s, 'the environment (improvement) costs my business money'. Several estimates of waste costs to a business are in the region of 4 per cent of turnover. Alongside better energy management, waste is a significant area of financial savings for a business.

Organisational approaches to achieving zero waste

It is perhaps within organisations that the opportunity to achieve a zero-waste operation is most likely to be achieved. The reason for this is that, unlike domestic situations, organisations have complete control (or should have) over their processes, operations and staff. In addition, organisations tend to have a restricted range of wastes, generally those specifically associated with the industrial sector in which they work, though there may be small quantities of other waste, and hence the objective of achieving zero waste will be more easily achieved.

Therefore, within an organisation trying to achieve zero waste, the following management strategies need to be established.

Management commitment

This does not mean that senior management will verbally give support, but that they will be prepared through their own actions to demonstrate zero-waste performance. They will only take decisions which result in zero waste, they will only invest in 'waste-free' products and services, and they will show 'zero tolerance' of wasteful performance by their staff.

Communication

Once zero waste is established as an organisational objective, it will be vital to communicate this to all the stakeholders of the organisation. The frequent problem with communication is that if recipients do not understand the message, do not agree with the message, or do not recognise the relevance of the message to their work, that message will largely be wasted and will need support and reinforcement.

Set out below in Table 1 is a matrix offering the opportunity to examine the communication strategy appropriate to different stakeholders:

Table 1. Communication strategies and stakeholders

Stakeholders' communication strategy	Middle and department managers	Staff	Suppliers	Customers and clients	Waste contractor	Board	Partners
Workshops							
Emails							
Face-to-face meetings							
Brochures and leaflets							
Notices/posters							
Press releases							
Waste mountains: 'piles' of waste in reception in restaurants							
Waste tips to reduce, reuse recycle on the intranet, pay slips, contracts							
Zero waste events such as meetings, conferences, etc.							

Incentives and penalties

The use of incentives and penalties may, depending on the particular organisational culture, play an important role as reinforcement for other strategies. Typically, when organisations proceed to eliminate waste from their processes, it is undoubtedly the organisation that benefits! The elimination of waste will impose significant strategic challenges for the organisation, and, if successful, it will result in major financial and other working benefits both short and long term. In the past, such benefits were typically not shared with the employees, with little in the way of results achieved (see below). Today, the question is whether the sharing is financial or through staff benefits (Box 5).

Box 5. To share or not to share?

In one London borough about 10 years ago, officers were concerned with the energy used in the large civic centre, and in conjunction with CESMB, they developed a plan to provide staff who saved energy with an incentive of about 50 per cent of the cost of the energy saved. However, when this plan was put before the council, the reported response was, 'Well, if they can make changes which save energy, they can do it without an incentive.' Needless to say, the energy consumption in the civic centre continues to be high.

Governance and structures

Many organisations find that when tackling a waste problem that runs through the organisation – and waste certainly is an organisation-wide issue – it is very helpful, in implementing the zero-waste strategy, to develop a cross-departmental team who themselves develop an ethic, a culture and an expertise to support different areas of the organisation in achieving zero waste. In organisations with different departments and/or different sites, the comparison between the successes of different sites or departments in achieving waste elimination can show up areas of success and areas of difficulty (Box 6).

Box 6. Implementation

Middlesex University has several sites scattered throughout north London, and, until recently, it had an environmental advisory group. One of the objectives of this group was to encourage each campus to reduce waste as much as possible, and largely this meant recycling as much paper as possible. It became very noticeable that one campus was consistently under-performing; in fact, not performing at all. When an opportunity to make staff changes allowed a key member of the caretaking staff to move from a good-performing site to the poor performer, the change was dramatic.

Investment

Any organisation proposing to achieve zero waste needs to demonstrate management commitment. This means senior management must perform at least to the same levels and standards that they expect from their staff. It also means management investment in changes, technologies, training and processes which will underpin and support the move towards zero waste.

Communication as a key issue in achieving zero waste

All the organisational strategies outlined above must be communicated clearly to the stakeholders. Several communication strategies are clearly relevant to the achievement of zero waste, and each of these strategies must recognise the position of the stakeholder.

There probably must also be strategies to communicate to staff the changes in performance that they must make to achieve zero waste, and to tell them how to use new equipment correctly, and this must be repeated until the new practices become accepted (and are seen as the culture) within the organisation. This change of culture must be supported not just by effective messages but also by repeating them again and again.

The communication with staff in the organisation needs to recognise that as part of that communication the need for staff to develop their awareness and understanding of waste management is integral. It would be unlikely to achieve much and would be unsustainable if the communication strategy for staff did not appreciate that appropriate training techniques are an important ingredient of the communication strategy. Modern approaches to blended training, in which software, paper-based and workshop approaches are all integrated into a comprehensive, work-based and cost-effective training programme, are likely to be effective in achieving a more lasting approach to zero waste (Box 7).

Box 7. A blended training approach

CESMB at Middlesex University have developed a blended training approach to waste management which incorporates five workbooks, a CD and a website on which participants can use the materials but also develop links and establish a forum for discussion with other participants. The programme is also augmented with briefing sessions and masterclasses.

The programme at this stage does not extend to taking participants through to a zero-waste strategy. Probably 90 per cent of participants have been involved in waste management in their organisation only to the extent of throwing waste in the bins, and the concept of waste reduction, reuse, etc., is fairly novel. However, once they have embarked on the programme, its interactive approach will lead many of them to begin to identify the potential that a zero-waste strategy might offer.

Another major opportunity for communicating an organisation's aim to achieve zero waste could be with their supply chain. Such is the dependence of modern processes and organisations on an extensive supply chain that without integrating the supply chain within the zero-waste aim, it is highly unlikely that significant results will be achieved. The communication strategy to be adopted with suppliers might follow one of three different scenarios:

1. Tell the suppliers what is required in terms of waste reduction and reuse of materials and equipment, and rely on them to supply to your specification.

2. Work with the suppliers to develop a joint strategy leading to zero waste.

3. Collaborate with the suppliers in the development of new designs and products using new materials that will lead to a more sustainable zero-waste strategy.

Any of these approaches is likely to lead to the elimination of waste, but, clearly, the level of communication, joint working, training and development in strategy 3 and, to some extent, strategy 2 is much more likely to result in a sustainable approach to zero-waste management (Box 8).

Box 8. The 10 steps to zero waste for an organisation

1. Assess the types and quantity of waste currently produced.

2. Track back each type of waste to its source.

3. Identify an alternative material or equipment that produces less waste.

4. Examine the processes that the resources go through before the waste is created.

5. Examine how the processes can be changed to create less waste.

6. Follow through the 'reduction, reuse, remanufacture/redesign, recycle, treat' waste hierarchy for every item in your organisation.

7. Start an effective zero-waste collaboration with your suppliers.

8. Preferentially purchase remanufactured or recycled goods.

9. Examine the corporate strategy of your organisation and develop it to meet a zero-waste strategy.

10. Communicate all this to your staff and train them to perform to different standards.

Conclusion

Zero waste is an achievable dream and would have significant financial, environmental and social benefits. But, unlike most dreams, zero waste needs considerable effort to make it sustainable. Critical in that objective is the strategy of communication, but that, in turn, must be supplemented with appropriate corporate management strategies. Communication strategies need to be viewed within a very broad context, and they include aspects such as training, electronic methods, and events and exhibitions.

Further References

Anthony, S. *'The Environment' – What Every Business Needs to Know* (London: Middlesex University Press, 2005).

Goodland, R. *Sustainability: Human, Social, Economic and Environmental* (Wiley, 2002).

Louis, R.S. *Creating the Ultimate Lean Office – A Zero Waste Environment with Process Automation* (New York: Productivity Press, 2007).

Murray, R. *Creating Wealth from Waste* (London: Demos, 1999).

Websites

www.wasteonline.org.uk
www.greenpeace.co.uk
www.environmentcity.org.uk
www.ippr.org/publications&reports
www.dove2000.org.uk

Contributor profiles

Editors

Dr Kenny Tang CFA is founder chief executive officer of WASTEnomics Capital and Oxbridge Weather Capital, leading experts in the waste, weather, low carbon, cleantech and climate change space. Kenny has postgraduate degrees from the Universities of Oxford and Cambridge. Dubbed as 'Asia's Al Gore' by leading global investment bank Merrill Lynch and global strategy magazine *Strategic Direction,* Kenny has written on sustainability, climate change, cleantech, waste and green entrepreneurship for the *Financial Times* and *Wall Street Journal.* He is on the board of governors at the University of East London (having completed two years as governor at Middlesex University) and also a visiting fellow/adjunct professor teaching on the world's first MBA in strategic carbon management at the Norwich Business School (University of East Anglia).

He sits on the global judging panel of the *Wall Street Journal's* Technology Innovation Awards and the *Asian Wall Street Journal's* Asian Innovation Awards. He is a Chartered Financial Analyst (CFA) charter holder from the CFA Institute. He authored the first book on financing university spinouts: *'Taking Research to Market – How to Build and Invest in Successful University Spinouts'* (Euromoney Books, 2004). His acclaimed book: *'The Finance of Climate Change – A Guide for Governments, Corporations and Investors'* was published to coincide with the UK presidency of the G8 Group of Nations and the Gleneagles Summit in July 2005 and received a special foreword from UK Prime Minister Tony Blair. His latest book on the corporate response to climate change (with Ruth Yeoh): *'Cut Carbon, Grow Profits – Business Strategies for Managing Climate Change and Sustainability'* was launched at YTL's Climate Change Week in Kuala Lumpur in March 2007.

Jacob Yeoh is an executive director in the construction division of YTL Corporation, a US$9bn corporation in the power, water, retail, hotel, leisure resorts, property, construction, cement and e-solutions sectors. Before that he worked in UBS in the wealth management department in Singapore. He graduated with a Masters degree in electrical and electronic engineering at Imperial College, University of London. His research interests are in renewable energy markets, construction, commodities, clean technology and investments.

Contributors

A. Prem Ananth is a research associate working with the 3R Knowledge Hub at the Asian Institute of Technology, Bangkok, Thailand. His current areas of interest include solid waste management focusing on 3R, industrial ecology, CDM and climate change and energy–environment interactions.

Stewart Anthony is head of the Centre for Environment and Safety Management for Business (CESMB) training and research Centre within Middlesex University. He has been involved as trainer, adviser and collaborator with organisations, businesses and social enterprises in the field of sustainable development and environmental management since 1998. Many of the projects he has run have been EU or government funded and in the past 15 years has worked with over 500 organisations in the UK and Europe. He is the author of *The Environment – what every business needs to know*.

Ranjit Singh Baxi BSc, MBA, FRSA established his recycling business, J and H Sales (International) Ltd. 25 years ago, and is its chairman. The company exports secondary fibre from the UK, Europe and the USA to many Asian countries. Its success was recognised when it received the Queen's Award for Enterprise (International Trade), in 2001. R S Baxi is president of the paper division of the Bureau of International Recycling (BIR), with its head office in Brussels. He regularly speaks at European and International conferences. Mr Baxi is also a governor of the University of East London (UEL) and is a non-executive board director of Think London & Gateway to London, both government-sponsored organisations promoting inward investments into London.

Peter Beigl is a research associate at the Institute of Waste Management at the BOKU University of Natural Resources and Applied Life Sciences, Vienna since 2002. He finished his studies of civil engineering and water management at BOKU University in 2002, and business administration at the Vienna University for Economics and Business Administration in 2003. His main fields of interest are waste logistics, waste generation and waste prevention, collection and recycling of waste, life cycle analysis in waste management, ecodesign and waste prognosis modelling.

Paul Carey is technical and development director of the municipal infrastructure developments division for Waste Recycling Group, concentrating on energy from waste, mechanical biological treatment and other resource recovery solutions. Paul's experience spans large and small-scale fossil fuel and renewable energy generation, including energy from waste, with more than 15 years in the industry. He is presently involved in a number of major project developments for WRG, including a 240,000 tpa energy from waste facility at Hull.

Jane Fiona Cumming is an experienced strategic advisor to boards in public and private sector organisations around the CSR and sustainability agenda. She is also an effective implementer, particularly in issues around business responsibility, governance, emerging risk, futures, engagement, behaviour change and making the issues practical and intelligible to the layman. Originally a life scientist, Jane Fiona co-founded Article 13, the business responsibility experts (www.article13.com). Her first degree, in life sciences, is from Durham University, her MBA in strategic marketing is from Hull University and she has a Diploma in corporate governance (ACCA).

Tim Forsyth is reader in environment and development at the London School of Economics. He is the author of *'International Investment and Climate Change: Energy Technologies for Developing Countries'* (Earthscan, 1999), and numerous other articles about technology transfer and the climate change negotiations. He works more generally on topics of environmental governance and integrating local developmental concerns into global environmental policy, especially in Asia.

Sandra Greiner is a senior project manager at Climate Focus and an expert in developing CDM and JI projects. Sandra has extensive experience guiding energy and waste management projects through the CDM registration process, and is a leading expert in baseline methodology development. Before joining Climate Focus, Sandra was a carbon finance specialist at the World Bank. Sandra holds a PhD in economics from Hamburg University in Germany.

William Hogland graduated Master of Science in civil engineering at Lund University, Sweden. He became Doctor of Science in water resources engineering and some years later associate professor in the same area at Lund University. His activities at the University of Kalmar began during the second half of the 1990s when he became acting professor in technical environmental science where he later gained a professorship in environmental engineering and recovery. In 2007 he also gained a professorship in eco-technology with a focus on innovation systems at Mid Sweden University, Sweden. He has published more than 400 scientific papers and reports in the area of waste and water management and has experience of teaching and research in over 50 countries. He has been invited as a guest speaker at universities in 18 countries all over the world.

Jelmer Hoogzaad is a technical expert in CDM and JI project assessment, risk management, and green investment schemes. Before joining Climate Focus, he worked as a project officer for Russia, Ukraine and Bulgaria at ERUPT, the carbon credits procurement programme of the Dutch government. Jelmer holds an MSc in natural sciences and innovation management from Utrecht University in the Netherlands.

Kurian Joseph is an environmental engineer, currently assistant professor in environmental engineering at the Centre for Environmental Studies, Anna University, Chennai, India. He is an investigator on the Asian Regional Research Project on Sustainable Landfill Management in Asia, accredited environmental auditor, author of several technical papers and organizer of continuing education programs in emerging areas. His areas of interest include cleaner production, municipal solid waste management, industrial and hazardous waste treatment and environmental management. He received the 'Kriton Curi' award for the best paper in waste management from developing countries at the IX International Waste Management and Landfill Symposium in Sardinia, Italy, October 2003. He is an author of the book: *'Essentials of Environmental Studies'* published by Pearson Education.

Adriaan Korthuis is a founding partner and director of Climate Focus. He is an authority on the design and implementation of the Kyoto Protocol, with a particular focus on the Clean Development Mechanism (CDM) and Joint Implementation (JI). Prior to founding Climate Focus, he was initiator and programme manager of the ERUPT and CERUPT programmes of the Dutch government. Adriaan holds an MSc in Food Science from Wageningen University in the Netherlands.

Martin Kurdve is responsible for the development of waste management and chemical management for Volvo Group. He has a background as energy engineer at Forsmarksskolan and an MSc in Chemical Engineering with Engineering Physics from Chalmers University of Technology. After working as an IT consultant he started at Volvo Technological Development Corporation in 2000 as an environment and chemistry development engineer for production plants. He is a part-time PhD student at the Institute of International Industrial Environmental Economics at Lund University, researching chemical management and product service systems.

Erik Laes obtained a degree in chemical engineering at the University of Leuven in 1998. He also holds degrees in philosophy (1997) and environmental studies (1999). In 2006, he finished his PhD project on the sustainable energy debate in Belgium and the role of nuclear power. He is currently involved in research on stakeholder involvement in radioactive waste governance.

Sandra Lebersorger is a research associate at the Institute of Waste Management, Department of Water, Atmosphere and Environment, BOKU – University of Natural Resources and Applied Life Sciences, Vienna, Austria. She has a degree in land and water management and engineering and finished her PhD thesis on 'Waste Generation from Multifamily Dwellings', for which she received an award from ISWA Austria in 2004. Her main fields of interest are waste logistics, waste generation in private households, social/psychological aspects and waste prevention.

Marcia Marques (Gomes) gained a professorship in eco-technology from Mid Sweden University – MIUN, Sweden (2008), Docent from the Department for Water Environment Transport, School of Civil Engineering, Chalmers University of Technology (2005), PhD in chemical engineering from the Royal Institute of Technology – KTH, Stockholm (2000), MSc in Biology (1980) and BSc cum laude in biology (1976) from the Federal University of Rio de Janeiro – UFRJ. She has worked as an international consultant for UNEP (2000–2004). Her main areas of interest are: bioremediation and phytoremediation of contaminated soils, biological processes for treatment of industrial effluents; technologies for pollution abatement based on biological systems and strategic environmental impact assessment.

Robert O'Sullivan was trained as a lawyer and specializes in regulatory and policy aspects of the CDM and JI. He is an expert in carbon finance transactions, both on the buyers' and sellers' sides of the negotiation table and was one of the lead drafters of the CERSPA initiative. He holds degrees in sciences, arts and law from the University of Queensland in Australia and a Masters of Law from The American University in Washington DC.

Adam David Read is a Chartered Waste Manager, and head of waste management at Hyder Consulting, driving forward their UK waste management consultancy portfolio of work, including the design and delivery of participatory planning and awareness raising programmes. For four years he was a member of the international waste team at ERM, where he led public engagement, stakeholder review and awareness raising activities for waste collection and recycling services in the EU, accession states and a number of developing economies. Adam was awarded his doctorate in solid waste management policy and planning in October 2001 from Kingston University in London, and prior to this he had been a recycling officer with the Royal Borough of Kensington and Chelsea.

Ella Stengler has been the managing director of CEWEP (Confederation of European Waste-to-Energy Plants) since July 2003. CEWEP represents 340 Waste-to-Energy plants from 17 countries. Previously she was the managing director of ITAD, the German Association of Waste-to-Energy plants, and prior to that she was the director of AGS, the German association of hazardous waste companies. Ella has published several articles in German and international magazines. She studied law in Germany and France. She wrote her doctoral thesis on the subject: recovery and disposal of waste according to national and Community law published in 2000. Since 2000 she is also registered as a lawyer.

Jan Stenis was conferred the title Doctor of Science/PhD in construction management having defended the thesis Industrial Management Models with Emphasis on Construction Waste in 2005. In 2002 he became Licentiate in Engineering on the dissertation Industrial Waste Management Models – A Theoretical Approach. He holds an MSc in civil engineering from Chalmers University of Technology, Sweden which includes a Masters course in management science from Imperial College of Science, Technology and Medicine, London. He has a BSc in business administration from the School of Economics and Management at Lund University, Sweden. Dr Stenis has some 20 years of qualified working experience with five years in leading positions with responsibility for technological, financial and/or administrative matters.

Charlotte Streck is a founding partner and director of Climate Focus and an international law expert in climate change and carbon finance law and policy. She advises on the regulatory framework of the United Nations Framework Convention on Climate Change (UNFCCC), the Kyoto Protocol, and the European Union Emissions Trading Scheme (EU ETS). Before she joined Climate Focus as director in February 2005, Charlotte was counsel at the World Bank in Washington.

Adewole Taiwo is the national co-ordinator of the Nigeria Climate Change Working Group (NCCWG) as adopted at the December 2007 Global Climate Change Day of Action in Lagos Nigeria. He is an environmental expert and a consultant on waste management, health and safety management and other environmental issues. Taiwo has an MSc in Environmental Resources Management and a BSc in Geography from Lagos State University, Nigeria. He is a member of the Nigeria Institute of Management, Nigerian Environmental Society and the International Solid Waste Association. He is the chief executive officer of Taiwo Adewole and Associates, an environmental consultant agency in Nigeria.

C. Visvanathan is a professor in the Environmental Engineering and Management Program at the Asian Institute of Technology, Bangkok. In addition to offering courses in environmental engineering and management, he also offers his consulting services to several multilateral organizations. His core areas of research and expertise include solid waste management, water and wastewater treatment, membrane technology, cleaner production and industrial environmental management.